AF361872

Challenge and Research Trends of Solar Concentrators

Challenge and Research Trends of Solar Concentrators

Editors

Dawei Liang
Changming Zhao

MDPI • Basel • Beijing • Wuhan • Barcelona • Belgrade • Manchester • Tokyo • Cluj • Tianjin

Editors
Dawei Liang
New University of Lisbon
Portugal

Changming Zhao
Beijing Institute of Technology
China

Editorial Office
MDPI
St. Alban-Anlage 66
4052 Basel, Switzerland

This is a reprint of articles from the Special Issue published online in the open access journal *Energies* (ISSN 1996-1073) (available at: https://www.mdpi.com/journal/energies/special_issues/ Solar_Concentrators).

For citation purposes, cite each article independently as indicated on the article page online and as indicated below:

LastName, A.A.; LastName, B.B.; LastName, C.C. Article Title. *Journal Name* **Year**, *Volume Number*, Page Range.

ISBN 978-3-0365-6037-3 (Hbk)
ISBN 978-3-0365-6038-0 (PDF)

Cover image courtesy of Dário Garcia, Dawei Liang, Cláudia R. Vistas, Hugo Costa, Miguel Catela, Bruno D. Tibúrcio and Joana Almeida

Contents

About the Editors

Dawei Liang

Dawei Liang is an Associate Professor of New University of Lisbon. He has published more than 130 articles. His researches on solar lasers were highlighted by CSP Today, Spotlights on Optics in 2012, Laser Physics in 2013, Laser Focus World in 2013, 2016 and 2022, Journal of Photonics for Energy in 2020, 2021 and featured news from SPIE in 2022. His research team signed a book-publishing contract on Solar-Pumped Lasers with Springer-Nature in 2020. He was among "World's Top 2% Scientists list" by Stanford University in 2021. He is an Associate Editor of Journal of Photonics for Energy. He serves as special issue Guest Editor for Energies.

Changming Zhao

Changming Zhao is a Full Professor of Beijing Institute of Technology. He was a visiting scholar of Technical University of Berlin. He has published more than 140 articles. He accomplished many research projects like LD-pumped double-frequency solid-state laser, LD-pumped Yb:YAG laser, DPL coherent laser radar techniques, Solar-pumped solid-state lasers etc. He serves as Editors of Lasers and Infrared (in Chinese) and Applied Optics (in Chinese). He published the first book on Solar-Pumped Lasers (in Chinese) in 2016. He serves as special issue Guest Editor for Energies.

Preface to "Challenge and Research Trends of Solar Concentrators"

Challenge and research trends of primary and secondary solar concentrators are key issues for advanced solar energy research. The main purpose of the book is therefore to present some of the most recent developments related to solar concentrators. This book is designed to provide the readers a better understanding of solar concentrators and their impacts on solar-pumped lasers. The chapter order follows a logical flow of ideas from novel primary solar concentrators to novel secondary concentrators, concluding finally with the design details of novel solar-pumped lasers.

1. Novel primary solar concentrators for high solar flux applications

A novel three-dimensional elliptical-shaped Fresnel lens (ESFL) analytical model was presented to evaluate and maximize the solar energy concentration of Fresnel-lens-based solar concentrators.

2. Novel compound primary solar concentrator for pumping a solar laser

Significant numerical improvement in end-side-pumped solar laser collection efficiency and beam brightness was presented by combining a Fresnel lens and a modified parabolic mirror.

3. Novel fixed solar concentrator for illumination

A fixed fiber light guide system using concave outlet concentrators as its receiving unit was proposed. The absence of a tracking structure highlights this research.

4. Novel luminescent concentrator

The authors show that by shaping the Luminescent Solar Concentrators (LSCs) in the form of an elliptic array, its emission losses can be drastically reduced.

5. Conical secondary concentrator for thermal applications

In this study, thermal performance of a conical solar collector was assessed with a new design of concentric tube absorber and the results were compared to the existing circular tube absorber.

6. New progress in solar-pumped lasers by NOVA University of Lisbon

6.1. This study focuses on the influence of two secondary concentrators: an aspherical lens and a rectangular light guide. A Ce:Nd:YAG solar laser pumped through the rectangular fused silica light guide was experimentally investigated, attaining 40 W solar laser output power.

6.2. Most efficient solar laser emission from a single Ce:Nd:YAG rod
The utilization of a small diameter Ce:Nd:YAG rod was essential to significantly enhance solar laser efficiency, attaining 4.5% record solar-to-laser power conversion efficiency.

6.3. Novel solar concentrator design for the production of doughnut-shaped and top-hat solar laser beams. The first numerical simulations of doughnut-shaped and top-hat solar laser beam profiles were accomplished through both ZEMAX® and LASCAD® analysis.

6.4. Novel solar concentrator design for the emission of 5 kW-class TEM_{00} mode solar laser beams from one megawatt solar furnace

A novel multi-rod solar laser pumping concept was proposed to significantly improve the TEM_{00} mode solar laser power level and its beam brightness by novel zigzag beam merging technique.

In summary, this book may empower the readers with up-to-date approaches for designing and implementing the next generation solar concentrators, as well as solar-pumped lasers.

The book would not be possible without the valuable contributions of all authors. Sustained helps from MDPI editorial teams are highly appreciated.

Dawei Liang and Changming Zhao

Editors

Article

Elliptical-Shaped Fresnel Lens Design through Gaussian Source Distribution

Dário Garcia, Dawei Liang *, Joana Almeida, Bruno D. Tibúrcio, Hugo Costa, Miguel Catela and Cláudia R. Vistas

CEFITEC, Departamento de Física, Faculdade de Ciências e Tecnologia, Universidade NOVA de Lisboa, 2829-516 Caparica, Portugal; kongming.dario@gmail.com (D.G.); jla@fct.unl.pt (J.A.); brunotiburcio78@gmail.com (B.D.T.); hf.costa@campus.fct.unl.pt (H.C.); m.catela@campus.fct.unl.pt (M.C.); c.vistas@fct.unl.pt (C.R.V.)
* Correspondence: dl@fct.unl.pt

Abstract: A novel three-dimensional elliptical-shaped Fresnel lens (ESFL) analytical model is presented to evaluate and maximize the solar energy concentration of Fresnel-lens-based solar concentrators. AutoCAD, Zemax and Ansys software were used for the ESFL design, performance evaluation and temperature calculation, respectively. Contrary to the previous modeling processes, based on the edge-ray principle with an acceptance half-angle of $\pm 0.27°$ as the key defining parameter, the present model uses, instead, a Gaussian distribution to define the solar source in Zemax. The results were validated through the numerical analysis of published experimental data from a flat Fresnel lens. An in-depth study of the influence of several ESFL factors, such as focal length, arch height and aspect ratio, on its output performance is carried out. Moreover, the evaluation of the ESFL output performance as a function of the number/size of the grooves is also analyzed. Compared to the typical 1–16 grooves per millimeter reported in the previous literature, this mathematical parametric modeling allowed a substantial reduction in grooves/mm to 0.3–0.4, which may enable an easy mass production of ESFL. The concentrated solar distribution of the optimal ESFL configuration was then compared to that of the best flat Fresnel lens configuration, under the same focusing conditions. Due to the elliptical shape of the lens, the chromatic aberration effect was largely reduced, resulting in higher concentrated solar flux and temperature. Over 2360 K and 1360 K maximum temperatures were found for ESFL and flat Fresnel lenses, respectively, demonstrating the great potential of the three-dimensional curved-shaped Fresnel lens on renewable solar energy applications that require high concentrations of solar fluxes and temperatures.

Keywords: Fresnel lens; Gaussian source; groove number; solar flux; optical efficiency; full width at half maximum

Citation: Garcia, D.; Liang, D.; Almeida, J.; Tibúrcio, B.D.; Costa, H.; Catela, M.; Vistas, C.R. Elliptical-Shaped Fresnel Lens Design through Gaussian Source Distribution. *Energies* **2022**, *15*, 668. https://doi.org/10.3390/en15020668

Academic Editor: Tapas Mallick

Received: 24 November 2021
Accepted: 13 January 2022
Published: 17 January 2022

Publisher's Note: MDPI stays neutral with regard to jurisdictional claims in published maps and institutional affiliations.

1. Introduction

Optical concentration provides strong cost leverage for photovoltaic cells [1]. High concentrated photovoltaic technology uses relatively inexpensive optics, such as mirrors and lenses, to concentrate sunlight from a broad area into a much smaller area of active semiconductor cell and converts sunlight directly to electricity [2,3]. The Fresnel lens has been widely used in the concentrated photovoltaic field, with the advantages of simple structure, light weight, low cost, easy processing, etc. [4]. However, it has a limited concentration ratio due to its strong chromatic aberration.

To overcome this issue, non-flat Fresnel lenses have been widely studied by many researchers, proposing new modeling methods and configurations with the aim of increasing the solar energy concentration. The shaped Fresnel lens conception was initiated in 1977 by Cosby [5]. One year later, a patent was filed by O'Neill [6], while Kritchman published his finding [7]. Since then, many other researchers throughout the world have been proposing their own non-flat Fresnel lens models and modeling processes [8–13]. Currently, the most notable work in this field is a published book by Leutz et al. [14], who presented an in-depth study of the non-flat shaped Fresnel lens.

The modeling process of the shaped Fresnel lens follows Snell's law (or law of refraction). The half-angle subtended by the Sun is the key parameter to set the size of the refractive prism facets by using the edge-ray principle, which adjusts the number of grooves required for the concentrator. It dictates the absolute angle at which the solar rays arrive at the surface of the Earth globe. The Sun–Earth subtended acceptance half-angle of 0.27° can be determined by the mathematical relationship between the radial size of the Sun and its distance to the Earth [15–17]. It stands valid from a theoretical and mathematical perspective, being considered by many researchers in the field of solar energy concentration [7–12,14–23]. However, this angle could only be accepted if the Earth had no atmosphere. The sudden change from vacuum to Earth's atmosphere significantly alters the trajectory of solar rays due to the law of refraction. The solar rays pass through various atmospheric layers, such as the thermosphere, mesosphere, stratosphere, and troposphere. Each individual layer's gaseous composition has its own refractive index, influencing the refraction angle. Moreover, the half-angle subtended by the Sun can be different depending on geographical location, time zone, and local atmospheric conditions, such as the presence of clouds, humidity, sand particles and pollution in the troposphere. Therefore, within the Earth's atmosphere, the acceptance half-angle should be larger than 0.27°. These complex systems that influence the refraction of sunlight on Earth were neglected by most researchers in this field, who had adopted the half-angle of 0.27° as a standard for their works [7–12,14–24]. This oversimplification in the acceptance half-angle calculation paired with the edge-ray principle has set and cemented the number and size of grooves in the production of Fresnel lenses in the market [25].

There are two categories of simulation methods for evaluation of the concentration systems: one by ray tracing and the other by analytical approach [26,27]. The ray-tracing method simulates a vast number of discrete and well-defined rays of different wavelength, energy and traveling trajectory. Each ray has its own trajectory, which is influenced by the transmission, reflection, or refraction in a medium within its path. It is only numerically completed when the ray hits an absorbing detector or reaches an annihilation condition. This category has an accurate prediction of the flux distribution, but with the cost of high computing complexity, power and time. Several pre-existing reputable ray-tracing tools are available in the market, such as Zemax [28–30], LightTools [19,31–33] and Soltrace [34]. These software are reliable, robust, user friendly, easy to learn and allow repeatability, making them great tools to expand new fields of research.

The analytical method describes the flux density distribution through self-written complex computational codes and programs. This undoubtedly decreases significantly the computing time and has been widely adopted [8–10,12,21,22]. However, each research group has its own source code, which could be vastly different from that of another. In addition, there are no guarantees that the written code has been implemented correctly or that all the essential parameters and functions have been included. Due to all these uncertainties, the results could also be significantly different from each other.

The detailed description and analysis of the focusing image is the backbone of solar concentration research. The maximum attainable concentration and the flux distribution at the focal zone are some of the few characteristics that should be thoroughly explored and meticulously presented. To ensure the validity of the numerical or analytical results, it should be experimentally verified or compared to an existing experiment.

In this work, a parametric modeling procedure of a three-dimensional elliptical-shaped Fresnel lens (ESFL) is presented, using Snell's law. The ESFL is designed in AutoCAD, and then imported into Zemax nonsequential ray-tracing software for numerical evaluation of the ESFL output performance. In this evaluation, a Gaussian distribution based on measured data from the literature [35] is considered to define the solar source, instead of the classical edge-ray principle method with a solar acceptance half-angle of 0.27°. To validate the analytical model, this was applied in the performance evaluation of a flat

Fresnel lens with published experimental data [36], whose results were in accordance with those obtained through the proposed model. An in-depth study of the influence of the ESFL parameters, such as focal length, arch height and aspect ratio on its output performance, in terms of concentrated solar flux, optical efficiency and full width at half maximum (FWHM), is also carried out. Furthermore, to the best of our knowledge, this is the first time that the ESFL concentration efficacy is evaluated as a function of the number of grooves, with optimal 0.4–0.3 groove numbers per millimeter being found. This study reveals that the number of grooves necessary to maximize the concentrated solar flux could be significantly reduced in relation to that reported by the literature and market, which may enhance the cost efficiency of the manufacturing process of Fresnel lens solar concentrators. From the abovementioned studies, the best ESFL design is found. The optimal concentrated solar flux value within its focal cone is then analyzed and compared to that of a flat Fresnel lens, which proved the effectiveness of the ESFL in substantially reducing the chromatic aberration and, consequently, on maximizing the solar flux. The temperature analysis of both concentrators is also performed by Ansys. The maximum temperatures of 2360 K and 1360 K are attained from the ESFL and the flat Fresnel lens, respectively, demonstrating once more the potential of the ESFL in many solar energy research and applications that require high solar flux and temperature.

2. Modeling of an Elliptical-Shaped Fresnel Lens

The ESFL model has an elliptical-shaped arch with a smooth surface facing the Sun and grooves on the opposite side, as shown in Figure 1. This guarantees the easier cleaning of the lens in case of dirt, without damaging the grooves.

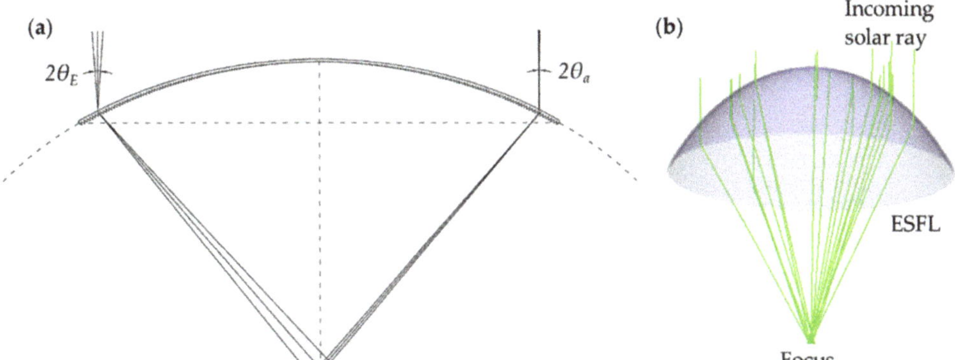

Figure 1. (**a**) Cross-sectional representation of an elliptical-shaped Fresnel lens (ESFL). $2\theta a$ represents the subtended angle of the Sun to the Earth and $2\theta_E$ the solar–terrestrial angle after the passage of the solar rays through the atmosphere. (**b**) 3D representation of the ESFL.

2.1. Analytical Method

The design of the ESFL followed a generic ellipse Equation (1), represented in Figure 2.

$$\frac{x^2}{a^2} + \frac{y^2}{b^2} = 1, \tag{1}$$

As shown in Figure 2, the conception of the elliptical arch depends on the radius or aperture of the concentrator (r), its focal height (h_f) and the height of the arch (h_l). The angle ω is the aperture angle or rim angle of the concentrator, with h_f and r defining its size. The ellipse's major axis "a" and minor axis "b" can be both calculated with those variables through Equation (2). The minor axis "b" is the combination of h_f and h_l and, once it is found, the major axis "a" can be acquired.

$$\begin{cases} a = \dfrac{b \times r}{\sqrt{b^2 - h_f^2}} \\ b = h_f + h_l \end{cases}, \tag{2}$$

The ESFL can be considered as a set of well-defined and positioned prisms, as shown in Figure 3.

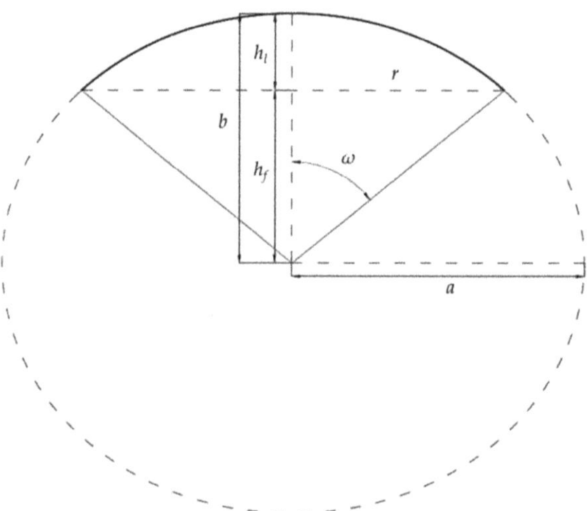

Figure 2. Segment of an ellipse used as an outline for ESFL. "*a*" and "*b*" are the major and minor axis of the ellipse. "*b*" is the sum of the height of the arch (h_l) and the focal length (h_f) and ω is the rim angle of the Fresnel lens.

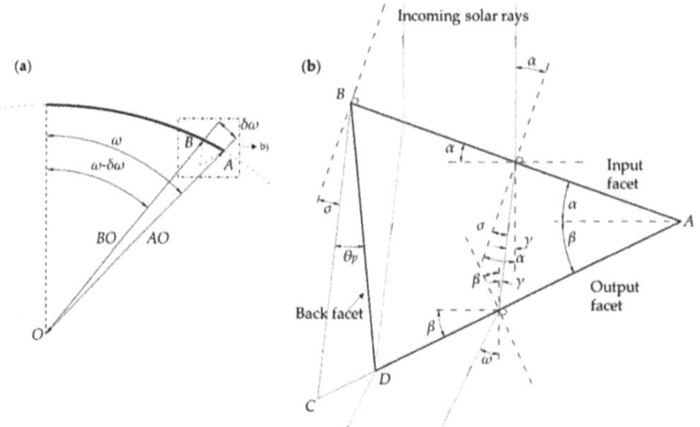

Figure 3. (a) Representation of a half-segment of an ESFL. The region defined by point A to B represents the first segment of the ellipse and the input facet of the first prism. The size of the prism is defined by angle $\delta\omega$. The point O represents the origin point and focal zone. AO and BO are the distances between the focal zone with the vertexes A and B of the prism, respectively. (b) Detailed visualization of a single prism. The incoming solar rays are refracted by the prism, outputting an angle ω in relation to the normal of its output facet, thus guaranteeing the concentration of light to the focal zone. AB segment is the input facet. Both AC and AD segments represent the output facet, whereas both BC and BD represent the back facet. The pitch angle (θ_p) defines the unused part of the prism.

The prisms were modeled through Equation (3), based on Snell's law, starting with the outmost one, as indicated in Figure 3a. Each prism is defined by three facets: the input, output and back facets, represented in Figure 3b. The input facet is defined by the segment between point A and point B. α represents the angle between the vertical line (solar ray path), with the normal line of the input facet AB. The solar ray is refracted as it hits AB. Depending on the refractive index of the medium (n_λ), the ray is deviated by an angle γ. It is then shifted further as its leaves the output facet of the prism, making an angle of ω in relation to the vertical axis. The output facet is either defined by the segment AC or AD depending on the pitch angle. In this paper, the refractive index n_λ of poly (methyl methacrylate) (PMMA) material (n_λ = 1.492) was considered.

$$\begin{cases} \sin(\alpha) = n_\lambda \sin(\alpha - \gamma) \\ n_\lambda \sin(\beta + \gamma) = \sin(\beta + \omega) \end{cases}, \tag{3}$$

The number of prisms/grooves (N) was calculated through Equation (4), where $\delta\omega$ is the groove division angle, which represents a small angular segment of the elliptical Fresnel lens.

$$N = \frac{\omega}{\delta\omega}, \tag{4}$$

Equations (5) and (6) give the distances of point A and B to the focal point (O, as the origin point), respectively, as shown in Figure 3a. The size of the input facet AB depends on $\delta\omega$. A larger $\delta\omega$ means a larger input facet, thus a larger prism. n is the current groove number.

$$AO_n = \frac{a \times b}{\sqrt{a^2 \cos(\omega - (n-1)\delta\omega)^2 + b^2 \sin(\omega - (n-1)\delta\omega)^2}}, \tag{5}$$

$$BO_n = \frac{a \times b}{\sqrt{a^2 \cos(\omega - (n)\delta\omega)^2 + b^2 \sin(\omega - (n)\delta\omega)^2}} \tag{6}$$

Knowing the distances AO_n and BO_n and the angle ω, then all coordinates can be found with basic trigonometry. The Cartesian coordinates of points A and B are then obtained through Equations (7) and (8), respectively.

$$\begin{cases} x_{A,\,n} = AO_n \sin(\omega - (n-1)\,\delta\omega) \\ y_{A,\,n} = AO_n \cos(\omega - (n-1)\,\delta\omega) \end{cases}, \tag{7}$$

$$\begin{cases} x_{B,\,n} = BO_n \sin(\omega - (n)\,\delta\omega) \\ y_{B,n} = BO_n \cos(\omega - (n)\,\delta\omega) \end{cases}, \tag{8}$$

The length of the facet AB at any given prism number is calculated by Equation (9):

$$AB_n = \sqrt{(x_{B,n} - x_{A,n})^2 + (y_{B,n} - y_{A,n})^2}, \tag{9}$$

The input angle α at any given prism number is given by Equation (10):

$$\alpha_n = \tan^{-1}\left(\frac{y_{A,\,n} - y_{B,n}}{x_{A,n} - x_{B,n}}\right), \tag{10}$$

The angle γ is obtained by isolating it from the remaining parameters of Snell's Equation (3), as shown in Equation (11):

$$\gamma_n = \alpha_n - \sin^{-1}\left(\frac{\sin(\alpha_n)}{n_\lambda}\right), \tag{11}$$

The calculation of angle β_n is acquired by directly applying Snell's law into Equation (10), resulting in Equation (12):

$$\beta_n = \tan^{-1}\left(\frac{n_\lambda \, \sin(|\gamma_n|) - \sin(\omega_n)}{\cos(\omega_n) - n_\lambda \, \cos(|\gamma_n|)}\right),\tag{12}$$

The exiting angle ω_n on the nth prism can be calculated through Equation (13):

$$\omega_n = \omega - (n-1)\delta\omega,\tag{13}$$

The inclusion of the pitch angle (θ_p) changes the length of the output facet from AC to AD. Equation (14) represents the calculation of the coordinates of point D considering an isolated prism, where point A is defined as the origin. This facilitates the calculation of the real coordinates of point D. The output facet length (AD) is then found by Equation (15).

$$\begin{cases} x'_{D,n} = \dfrac{\tan(90° - \theta_p - |\alpha_n| + |\gamma_n|) \, AB_n}{\tan(|\alpha_n| + |\beta_n|) + \tan(90° - \theta_p - |\alpha_n| + |\gamma_n|)} \\[2mm] y'_{D,n} = \tan(|\alpha_n| + |\beta_n|) \, x'_{D,n} \end{cases},\tag{14}$$

$$AD_n = \sqrt{{x'_{D,n}}^2 + {y'_{D,n}}^2},\tag{15}$$

The coordinates of the reduced output facet AD due to the pitch angle is represented in Equation (16).

$$\begin{cases} x_{D,n} = x_{A,n} - AD_n \cos(\beta_n) \\ y_{D,n} = y_{A,n} - AD_n \sin(\beta_n) \end{cases},\tag{16}$$

Finally, a single prism can be drafted with Equations (7), (8) and (16), forming a triangle of points ABD. The grooves of the ESFL's output surface were the first components to be modeled by drawing all the facets from coordinates A_n and D_n, starting from the outmost prism ($n = 1$) to the innermost prism N. This chaining process followed a sequence of $A_1D_1A_2D_2 \ldots A_ND_N$. The input surface of the ESFL was then drawn from the last prism (Nth prism) to the first one, i.e., N to $n = 1$, with the Y coordinate offset by an increment of d_t, the thickness of the concentrator from point A_N. Hence, the coordinates of point A_N take the form of ($x_{A,N}$, $y_{A,N} + d_t$). The output drawing is shown in Figure 4, and the procedural chaining process is summarized in a flowchart, as presented in Figure 5.

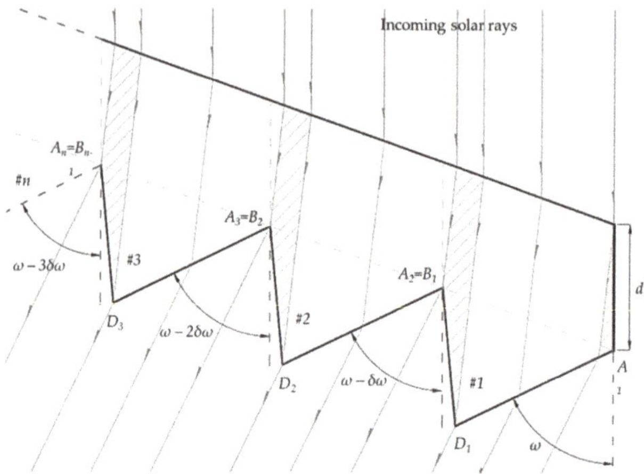

Figure 4. Chaining process of the first three prisms of the ESFL. The thickness of the concentrator is attributed by offsetting the input facet by d_t from its original coordinate position.

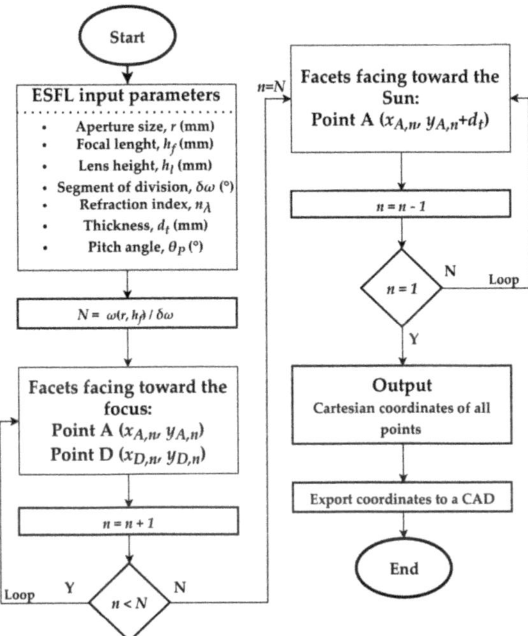

Figure 5. Flowchart of ESFL modeling process.

2.2. Numerical Simulation Method

Figure 6 shows the sequence of the numerical simulation process. The sequence was the same for each ESFL configuration, but each output was unique in terms of concentrated solar flux, optical efficiency and focal size at FWHM.

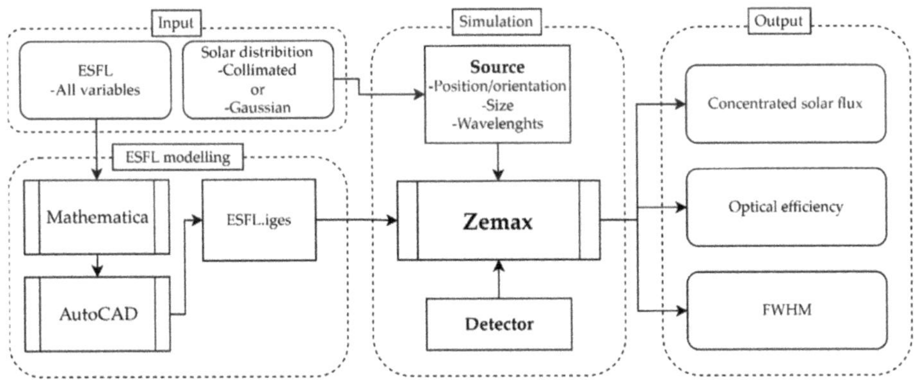

Figure 6. Sequence diagram of a single numerical simulation.

2.2.1. ESFL Modeling

Each individual coordinate of the ESFL was calculated with the Mathematica programming language. Then, the coordinates of all the relevant Cartesian points were exported into AutoCAD and linked as straight lines by the "polyline" command. The enclosed object was then revolved into a solid with the highest polygon count possible by adjusting the "facetres" command to its maximum. The object was then exported to Zemax as an *IGES* file. Both AutoCAD and Zemax shared the same coordinates, as shown in Figure 7.

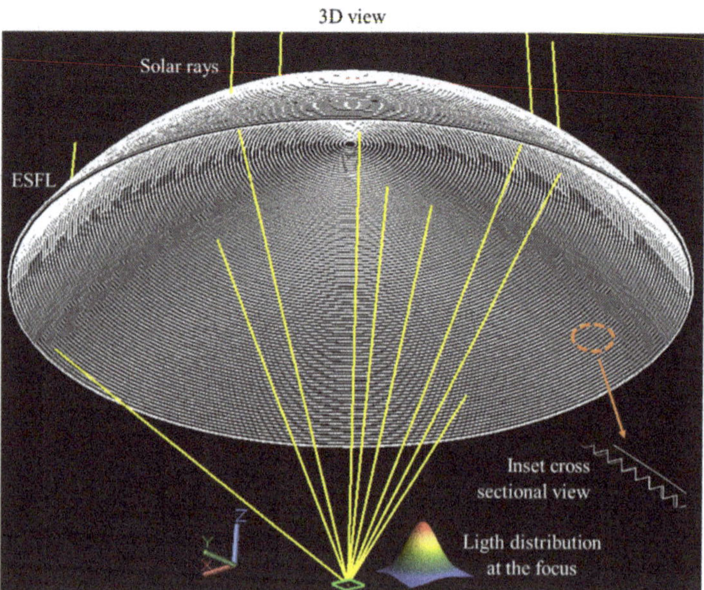

Figure 7. 3D view of the ESFL in AutoCAD.

2.2.2. Solar Source Modeling

An "ellipse source" with the same size of the concentrator was used to simulate the solar rays. The total power attributed is given by the product of the concentrator's collection area and the irradiance, i.e., a 1 m diameter concentrator would have a source power of about 785 W at 1000 W/m² irradiance. Figure 8 shows both the global and the direct reference spectra ASTM (American Society for Testing and Materials) G173 at Air Mass 1.5 (AM1.5) [37]. In Zemax, 21 wavelengths were selected as the solar spectrum data, each normalized as a function of its weight. The weight determines the intensity/power of each solar ray, which is dependent on the power attributed to the emitting source.

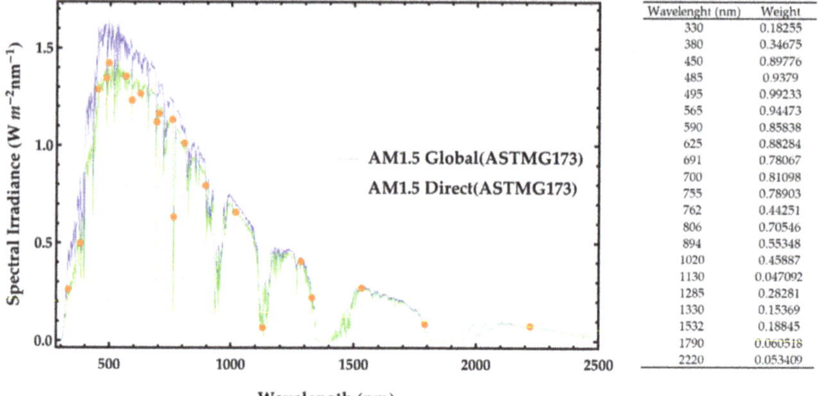

Wavelenght (nm)	Weight
330	0.18255
380	0.34675
450	0.89776
485	0.9379
495	0.99233
565	0.94473
590	0.85838
625	0.88284
691	0.78067
700	0.81098
755	0.78903
762	0.44251
806	0.70546
894	0.55348
1020	0.45887
1130	0.047092
1285	0.28281
1330	0.15369
1532	0.18845
1790	0.060518
2220	0.053409

Figure 8. Global and direct solar spectra ASTMG173 at AM1.5, with the selected wavelength used for ray tracing in Zemax and its respective weight value.

The Gaussian distribution of the solar source was defined by Zemax through the G_x and G_y parameters, i.e., the Gaussian distribution parameters in X and Y axes, respectively. These parameters are based on the solar irradiance measured by Vittitoe and Biggs [35].

The root-mean-square width (δ_{RMS}) adopted by Vittitoe and Biggs, as shown in Equation (17), describes the Sun shape as a function of the irradiance (I).

$$\delta_{RMS} \times 10^3 \approx 3.7648 - 0.0038413(I - 1000) + 1.5923 \times 10^{-5}(I - 1000)^2, \qquad (17)$$

Equation (18) represents the Gaussian distribution of the solar rays from the object "elliptical source" in Zemax. At 1000 W/m² solar irradiance, the Gaussian source would have a value of 35,276.6 at both G_x and G_y axes.

$$G_x = G_y = \frac{1}{2 \times \delta_{RMS}^2}, \qquad (18)$$

Alternatively, the Gaussian distribution of G_x and G_y can be manually adjusted in Zemax based on experimental measurements. For example, in a numerical study of a three-dimensional ring-array concentrator [28], where its output performance was compared to that of the medium sized solar furnace (MSSF) concentrator of PROMES-CNRS (Procédés Matériaux et Energie Solaire—Centre National de la Recherche Scientifique) [38], $G_x = G_y = 36,000$ was considered [28]. In this case, the solar distribution at the focal zone of the MSSF concentrator had the same characteristics as those described by [38–40].

The terrestrial solar half-angle (θ_E) can be calculated by determining the effective size of the solar–terrestrial image (d_E) at a distance (L), as represented in Equation (19).

$$\theta_E = \tan^{-1}\left(\frac{0.5\, d_E}{L}\right), \qquad (19)$$

θ_E can also be found in Zemax by using a detector at a certain distance from a small solar light source, as shown in Figure 9a. The length L is set as 10 m from the source of 0.002 mm diameter, and the size of the detector (d) as 150 × 150 mm² with a precision of 1001 × 1001 pixels. The calculation of θ_E is independent of the distance L since d_E is adjusted by the inverse-square law.

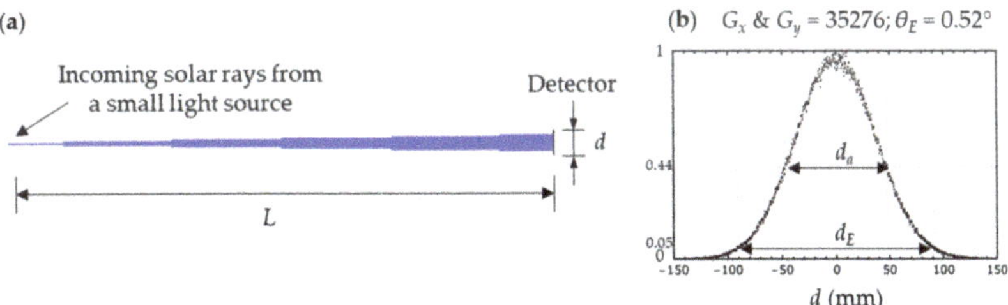

Figure 9. (a) Light propagation from a small light source towards a detector of size d at distance L. (b) Solar–terrestrial image with solar half-angle (θ_E) of 0.52°.

Figure 9b shows the solar distribution of the solar source obtained from both Equations (17) and (18), considering an irradiance of 1000 W/m² with $G_x = G_y = 35,276$. For an accurate calculation of the terrestrial solar angle, 95% of the total d_E distribution was considered and $\theta_E = 0.52°$ was calculated. It is important to note that the acceptance half-angle (θ_a) of 0.27° can be found at the effective size (d_a) by considering 56% of the focal Gaussian distribution.

2.2.3. Output Solar Distribution at the Focal Zone of the ESFL

To analyze the solar distribution at the focal zone of each ESFL configuration, over 60 million rays were employed per simulation. The output data were obtained from an absorbing detector with dimensions 20×20 mm^2 and resolution 150×150 pixels, positioned at the mathematical origin. The obtained data contained the solar distribution characteristics in terms of concentrated solar flux, optical efficiency (the total number of rays that strike the detector over the total emitted rays from the source) and FWHM.

Figure 10 shows the comparison of the three-dimensional focal distributions from a Gaussian source with $G_x = G_y = 35{,}276$ and a collimated source with $G_x = G_y = 0$ (or infinite). In this case, an arbitrary ESFL model with 1 m diameter (D), $h_f = 400$ mm, $h_l = 400$ mm, $d_t = 3$ mm, and $\delta\omega = 0.20°$ (256 grooves), with $\theta_p = 12°$, was used. The focal shape formed by the Gaussian source (Figure 9a) has a wide normal distribution with 10.8 mm FWHM and a concentrated solar flux of 5.0 W/mm^2, while the collimated source (Figure 9b) has a needle-shaped distribution with 0.3 mm FWHM, resulting in a peak concentrated solar flux of more than 100 W/mm^2. The collimated source offers the highest concentration intensity and its FWHM is nearly as tight as a laser beam. However, this is not possible to obtain with incoherent light provided from the Sun.

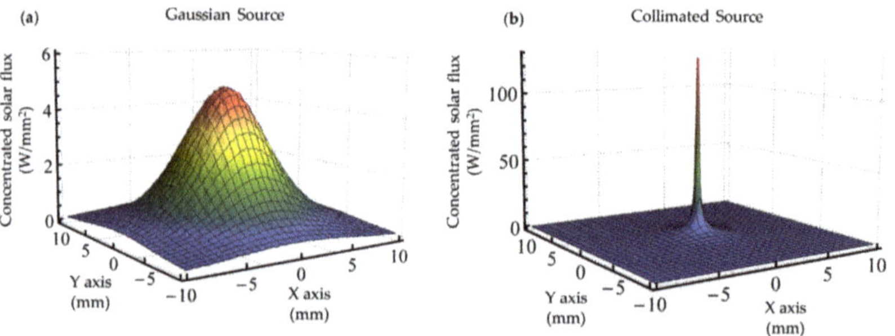

Figure 10. The focal image formed by ESFL modeled at 1000 W/mm^2 irradiance with the (**a**) Gaussian source and (**b**) collimated source.

2.2.4. Comparative Study of the ESFL Output Performance with the Measured Output Performance of a Fresnel Lens

Figure 11 shows the simulated layout and the focal output of the ESFL and a flat Fresnel lens [36] at the same focusing conditions. Ferriere et al. used a flat Fresnel lens of 900 mm diameter with 757 mm focal length, 20 grooves/cm (1800 grooves), and 31.7 mm thickness [36]. The measurement was conducted in PROMES-CNRS, where the irradiance can reach 1000 W/m^2 [36,38,41]. The remaining Fresnel lens parameters were adjusted to achieve the same output performance of the publication, by assuming a conic constant of -1.95 and a modest pitch angle of $2°$. By using the solar source Gaussian distribution from [35], the simulated focal output of the flat Fresnel lens in Zemax matched well with the measured data [36]. An ESFL with same collection size and focal length, and $h_l = 300$ mm, $\delta\omega = 0.0017°$ (1807 grooves), $d_t = 3$ mm and $\theta_p = 12°$, was numerically simulated and compared. Figure 11a,b present the cross-sectional view of the light rays from five concentric annulus solar sources (with same area and power) passing through the ESFL and the Ferriere Fresnel lens, respectively. Area 1 represents the solar rays from the outmost concentric annulus source, while Area 5 corresponds to the solar rays of the innermost circular source. Figure 11c,d show the contribution of each source on the concentrated solar flux of the ESFL and Ferriere Fresnel lens, respectively, as well as the combined concentrated solar fluxes distribution. As shown in Figure 11b, it is noticeable that the solar rays at the external annulus area of the flat Fresnel lens (Area 1 and 2) are barely detected, leading to solar fluxes close to zero (Figure 11d). The curved shaped of the ESFL overcomes this problem, allowing the solar rays from the external annulus area to be more efficiently focused, as demonstrated in Figure 11a,c. Since the external annulus areas collect a majority of the incoming solar power, the advantage of the ESFL concentrator in attaining higher solar flux becomes evident. Chromatic aberration is also significantly reduced, leading finally to a higher concentrated solar flux of 4.5 W/mm^2 (Figure 11c) compared to that of the flat Fresnel lens with 2.6 W/mm^2 (Figure 11d) [36]. In addition to the elliptical shape of the Fresnel lens, there are also important factors that can contribute to the better performance of this concentrator, such as the aspect ratio and the number/size of grooves. The influence of these factors on the ESFL output performance is addressed in Section 3.

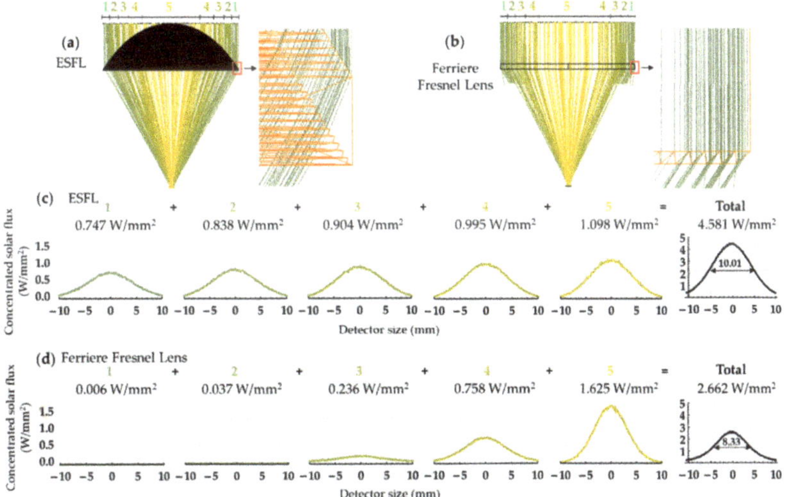

Figure 11. (**a**) The ESFL and (**b**) the Ferriere Fresnel lens irradiated by five concentric solar sources with identical area and power. (**c,d**) Respective concentrated solar fluxes for the ESFL and the Ferriere lens, respectively; the combined solar fluxes are also given.

2.2.5. Comparative Study of the ESFL Output Performance with Other Concentrators

Table 1 shows a summary of some predominant and documented analytical and measured data of other types of concentrators. It is important to note that the experimental data from the existing concentrators should be used as reference in the design and simulation of other concentrators, in order to obtain an accurate and fair evaluation of their performances.

The parabolic mirror of the MSSF had a measured peak concentration of 16 W/mm^2 with 2000 mm diameter, 850 mm focal length and a focal size of 10 mm FWHM [39]. This parabolic concentrator has currently the highest known concentrated solar flux. However, based on analytical predictions, some authors have claimed higher concentrations with focal width similar to that presented in Figure 10b [12,18,32,42], which is not possible with the incoherent solar light. The focal distribution measurements of linear-shaped Fresnel lenses by O'Neill and McDanal [6] and Leutz et al. [43] had both resulted in much lower concentration solar fluxes of 0.065 W/mm^2 and 0.045 W/mm^2, respectively, with the corresponding focal widths of 30 mm and 20 mm. It is common to find much lower concentrations from linear Fresnel lenses. The measured output performance of the flat Fresnel lens presented by Ferriere et al. had a concentrated solar flux of 2.6 W/mm^2 and a focal size of 8.3 mm at FWHM with a solar irradiance of 1000 W/m^2 [36]. The present ESFL, at the same conditions of the flat Fresnel lens from [36], had a better focal solar flux of 4.5 W/mm^2.

Table 1. Focal concentration and size from various concentrator types found in some publications.

Publications	Concentrator Type	Concentration	Focal Width (mm)	Lens Width (mm)	Focal Length (mm)	Grooves Per mm	Method
Nelson et al., 1975 [44]	Fresnel lens	4.3–5.0 (ratio)	~30	137	140	~0.1	Analytical
Cosby, 1977 [5]	Shaped Fresnel lens (linear)	~70 (ratio)	~20	~914	~509	~1.0	Analytical
Kritchman et al., 1979 [7]	Shaped Fresnel lens (linear)	172 (ratio)	N/A	vary	vary	Infinitely small grooves	Analytical
O'Neill et al., 1993 [6]	Shaped Fresnel lens (linear)	0.065 (W/mm^2)	~30	850	726	N/A	Experimental
Flamant et al., 1999 [39]	Parabolic mirror	16.0 (W/mm^2)	~16 ~10 (FWHM)	2000	850	N/A	Experimental
Leutz et al., 2000 [43]	Shaped Fresnel lens (linear)	0.045 (W/mm^2)	~20 ~5–6 (FWHM)	~300	~150	N/A	Experimental
Ferriere et al., 2004 [36]	Flat Fresnel lens	2.644 (W/mm^2)	~20 ~8.3 (FWHM)	900	757	2.0	Experimental
Yeh, 2009, [24]	Shaped Fresnel lens (linear)	0.060 (W/mm^2)	~5–6 (FWHM)	300	446	1.0	Analytical
Pan et al., 2011 [32]	Fresnel lens	1,367,704,600 (W/mm^2)	<0.25	189	189	N/A	Analytical
Akisawa et al., 2012 [10]	Shaped Fresnel lens	0.506 (W/mm^2)	<1	45	60	4.0	Analytical
Cheng et al., 2013b [18]	Fresnel lens	N/A	<0.1	88	50	0.22	Analytical
Languy et al., 2013 [12]	Shaped Fresnel lens	8500 (ratio)	<1	N/A	N/A	N/A	Analytical
Yeh, 2016 [21]	Shaped Fresnel lens (linear)	0.070 (W/mm^2) (at 1135 nm)	~20 ~4 (FWHM)	300	223	2.0	Analytical
Yeh et al., 2016 [22]	Shaped Fresnel lens	~5.0 (W/mm^2)	~20 ~4 (FWHM)	460	280	4.3	Analytical
Zhao et al., 2018 [33]	Shaped Fresnel lens (linear)	40.6 (ratio)	16	650	950	N/A	Experimental
Garcia et al., 2019 [28]	RAC [1]	16.0 (W/mm^2)	~20 ~10 (FWHM)	2000	500	N/A	Analytical
Liang et al., 2021 [42]	AFSCFL [2]	46.7 (W/mm^2)	<1	734	593	N/A	Analytical
Present work	Shaped Fresnel lens	4.5 (W/mm^2)	~10.0 (FWHM)	900	757	0.34	Analytical

[1] Ring array concentrator. [2] Annular Fresnel solar concentrator coupled with a circular Fresnel lens.

3. Output Performance of the ESFL Regarding the Concentrated Solar Flux, the Optical Efficiency and the FWHM

3.1. ESFL Configurations with Fixed Total Height of 700 mm

As any lens, the shape of the ESFL concentrator has direct influence on its output performance. Figure 12 shows some of the possible configurations of the ESFL with a combined h_f and h_l of 700 mm total height.

The ESFL has a similar configuration of a flat Fresnel lens at h_f = 700 mm, h_l = 0 mm. As h_l increases, the lens becomes more curved, and the aspect ratio becomes larger. The aspect ratio (AR) of the ESFL is given by the quotient of h_l and D (diameter of the lens):

$$AR = \frac{h_l}{D},\qquad(20)$$

Figure 13 shows the variation of the concentrated solar flux as a function of the aspect ratio of ESFL and h_l, with 700 mm total height. All the ESFL configurations had D = 1 m, $\delta_\omega = 0.28°$, $\theta_p = 12°$ and d_t = 3 mm. The concentrated solar flux starts with 3.9 W/m^2 at h_l = 0 mm, which closely resembles that of a flat Fresnel lens. It gradually increases with h_l, at which point the Fresnel lens is shaped into a parabola. The concentrated solar flux is maintained over 5.2 W/mm^2 from h_l of 500 mm to 700 mm, i.e., with aspect ratio from 0.603 to 0.702.

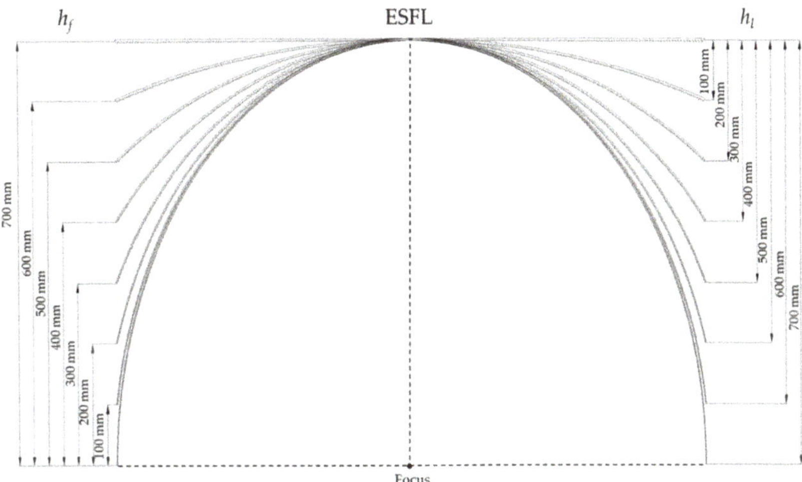

Figure 12. ESFL configuration with a combination of h_f and h_d for a total height of 700 mm.

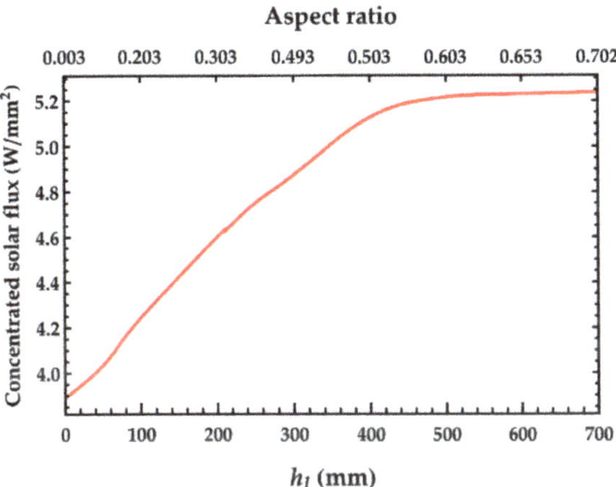

Figure 13. Concentrated solar flux of the ESFL with a combined $h_f + h_d$ of 700 mm as a function of h_l and aspect ratio.

3.2. Concentrated Solar Flux, Optical Efficiency and FWHM as a Function of the Aspect Ratio and Several Combinations of $h_f + h_l$

Figure 14 shows the concentrated solar flux, optical efficiency and FWHM at various combinations of h_f and h_l with ESFLs of $D = 1$ m, $\delta_\omega = 0.28°$, $\theta_p = 12°$ and $d_t = 3$ mm. h_f ranges from 200 mm to 600 mm and h_l ranges from 50 mm to at least 500 mm.

Each h_f has its own peak concentrated solar flux at different h_l (Figure 14a). For example, in the ESFL configuration with $h_f = 200$ mm, the concentrated solar flux varies from 4.2 W/mm² at $h_l = 50$ mm to 5.2 W/mm² at $h_l = 500$ mm, above which it starts to decrease. The highest concentrated solar flux was attained by the lowest $h_f = 200$ mm at $h_l = 500$ mm (aspect ratio of 0.603). The increase in h_f resulted in a lower concentrated solar flux, whose peak value shifted to a lower h_l, decreasing the aspect ratio of the concentrator. For example, the ESFL with $h_f = 600$ mm has an aspect ratio of 0.500 with $h_l = 350$ mm and a concentrated solar flux of 4.4 W/mm².

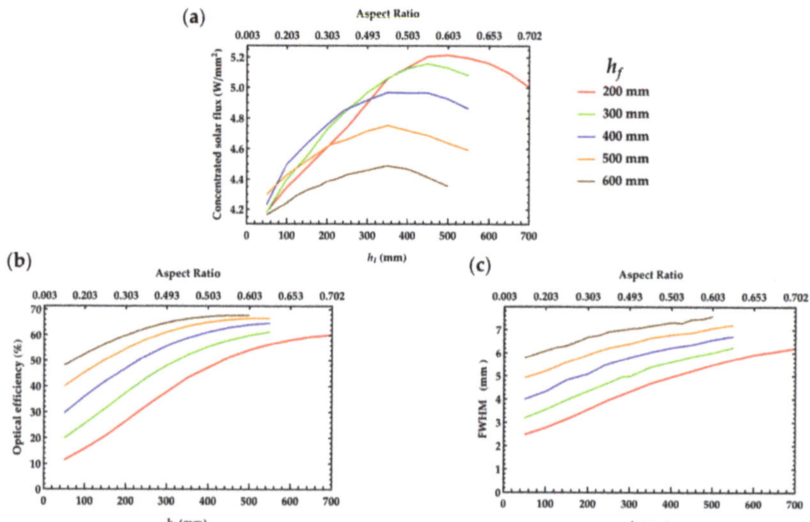

Figure 14. (**a**) Concentrated solar flux, (**b**) optical efficiency and (**c**) FWHM of the ESFL configurations with $D = 1$ m at various h_f and h_l combinations.

The optical efficiency tends to stagnate at larger h_l, as shown in Figure 14b. The highest optical efficiency of about 60% was attained at the higher h_f of 600 mm. The FWHM enlarges in a near linear fashion with the increase in h_l, and it also expands with the increase in h_f, as shown in Figure 14c.

3.3. Influence of Size/Number of Grooves on the ESFL Output Performance

The influence of different sizes of $\delta\omega$ on the ESFL output performance was analyzed for the most favorable combinations of h_f and h_l that achieved the highest concentrated solar fluxes (Figure 14a). The variation of $\delta\omega$ changes the size of each groove/prism within the ESFL concentrator, mainly changing the size of the input (A_nB_n) and the output (A_nD_n) facets, as calculated in Equations (9) and (15). A range of $\delta\omega$, from 0.05° to 1°, was used for the conception of the ESFL model. Figure 15a shows the number of grooves per $\delta\omega$ at different h_f. The number of grooves drops exponentially independently of the h_f with the increase in $\delta\omega$. A gap in the number of grooves between h_f is observed at a smaller $\delta\omega$, while it converges closer into the same number of grooves at a larger $\delta\omega$.

Figure 15b shows the concentrated solar flux as a function of $\delta\omega$. Regardless of h_f, the maximum concentrated solar flux was found at $\delta\omega = 0.3°$, which is equivalent to 210 to 132 grooves for h_f 200 mm to 600 mm, respectively. As shown in Figure 15c, the increase in $\delta\omega$ lowers the optical efficiency from 66% to 56% and 53% to 50% at h_f of 600 mm and 200 mm, respectively. The FWHM shows a minimum within a certain $\delta\omega$ range, as shown in Figure 15d. For example, at $h_f = 600$ mm, more than 10.7 mm FWHM was found at $\delta\omega = 0.35°$, while at $h_f = 200$ mm, the minimum FWHM of about 8 mm was found at $\delta\omega = 0.2°$. In all cases, the largest FWHM is located at $\delta\omega = 1°$.

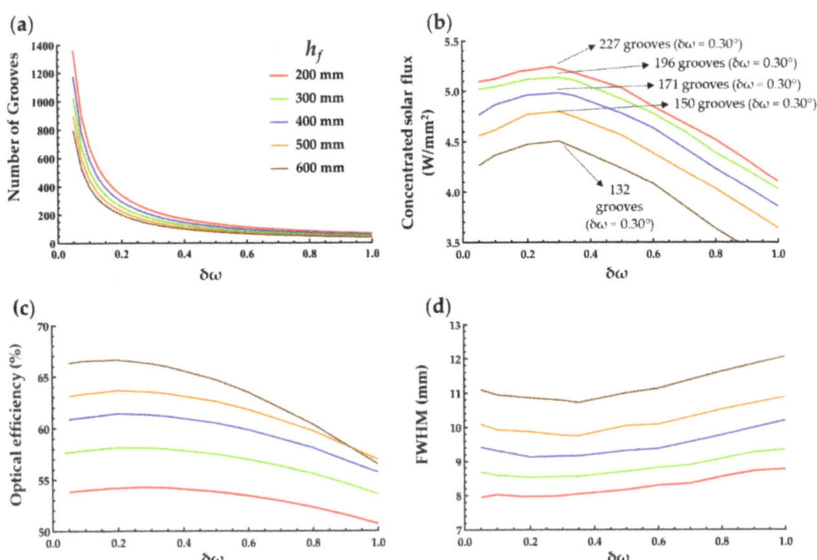

Figure 15. Characteristics of the Gaussian distribution source at the focal zone: (**a**) number of grooves per $\delta\omega$, (**b**) concentrated solar flux, (**c**) optical efficiency and (**d**) FWHM.

The number of grooves has a direct influence on the focal characteristics of the concentrator, as demonstrated in Figure 16. On the one hand, with the increasing number of grooves (lower $\delta\omega$), each prism becomes smaller, which reduces the refraction space within the prism. Consequently, above a certain number of grooves, the number of effective rays that could be refracted onto the targeted focal position is diminished, causing the decrease in the concentrated solar flux, as illustrated in Figure 16b. On the other hand, the reduction in the number of grooves (higher $\delta\omega$) leads to an increment in the size of the prism, hence broadening the final output focal shape, as shown in Figure 16d. Therefore, the concentration solar flux also diminishes with the decrease in the number of grooves (Figure 16b). Maximum concentrated solar flux was numerically found at $\delta\omega = 0.30°$ for all the ESFLs, regardless the h_f, as demonstrated in both Figures 15b and 16b. However, the optimum number of grooves varies with h_f. Consequently, for $\delta\omega = 0.30°$, the optimum number of grooves varied from 227 grooves with $h_f = 200$ mm to 132 grooves with $h_f = 600$ mm, corresponding to 0.42 grooves/mm and 0.26 grooves/mm, respectively. This variation is even more pronounced with smaller $\delta\omega$ (higher number of grooves). The optimum groove number per size is by far lower than that of other Fresnel lenses (Table 1), such as the shaped Fresnel lens with 4.3 groove/mm at $D = 460$ mm [22], flat Fresnel lens with 2 grooves/mm at $D = 900$ mm [36], 2 grooves/mm at $D = 889$ mm [45], and a range of 15.7 grooves/mm to 1 grooves/mm at various sizes of Fresnel lenses found at the market, such as the Fresneltech [25]. The higher performance of the ESFL with larger prisms and, consequently, lower number of grooves is attributed to the fact that the design process was not under the influence of the common modeling processes based on the edge-ray principle, but based on the solar Gaussian distribution. The reduction in the number of grooves and subsequent increase in the prism size could be favorable for the manufacturer in neglecting the limit losses due to manufacturing inaccuracies, such as blunt tips and deformed grooves forming.

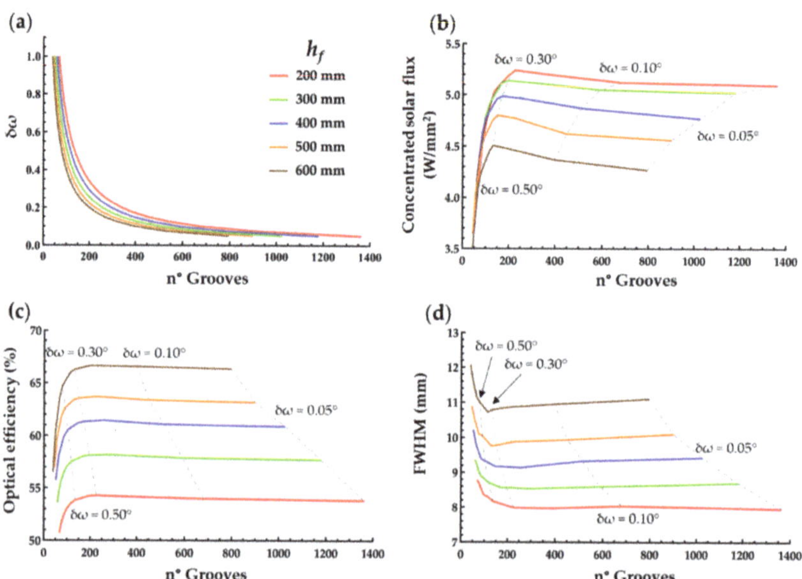

Figure 16. Characteristic of the Gaussian distribution source at the focal zone: (**a**) $\delta\omega$, (**b**) concentrated solar flux, (**c**) optical efficiency and (**d**) FWHM as a function of the number of grooves.

4. Comparison between ESFL and a Flat Fresnel Lens

4.1. Optimal Focal Position Analysis of Both the ESFL and a Flat Fresnel Lens

Figures 17 and 18 show the three-dimensional light distribution along the focal zone of the ESFL and a flat Fresnel lens with the same collection size, respectively. The ESFL had $D = 1$ m, $\delta_\omega = 0.28°$ (171 grooves), $\theta_p = 12°$, $d_t = 3$ mm, $h_f = 300$ mm, and $h_l = 450$ mm. The flat Fresnel lens with the highest concentration flux configuration was chosen, with $D = 1$ m, $h_f = 400$ mm, 0.3 grooves per mm (150 grooves) and $\theta_p = 12°$. The focal cones, as shown in Figures 17a and 18a, were represented by a detector volume with $100 \times 100 \times 100$ voxels in vacuum, where each voxel accumulates and stores the energy data from each ray that passes through it, with the associated wavelength and power. Figure 17b shows the top view focal distribution of the ESFL in the detector positioned at the origin ($Z = 0$ mm), with a concentrated solar flux of 5.08 W/mm², while Figure 17c shows the focal distribution at $Z = -5$ mm in relation to the origin, with a maximum concentrated solar flux of 5.48 W/mm². Figure 18b shows the focal distribution of the flat Fresnel lens at $Z = 0$ mm, with a concentrated solar flux of 0.75 W/mm². At $Z = +30$ mm in relation to the origin, the maximum concentrated solar flux of only 1.86 W/mm² was reached, as shown in Figure 18c.

For both the ESFL and the flat Fresnel lens, the focal distributions at the origin ($Z = 0$ mm) do not correspond to the positions where the concentrated solar flux is maximum, as shown in Figures 17a and 18a. This phenomenon occurs due to the chromatic aberration, which is more abundant in the flat Fresnel lens. It is important to note that a parabolic concentrator (a concentrator with no chromatic aberration effect) has its maximum concentration exactly at the origin point [28].

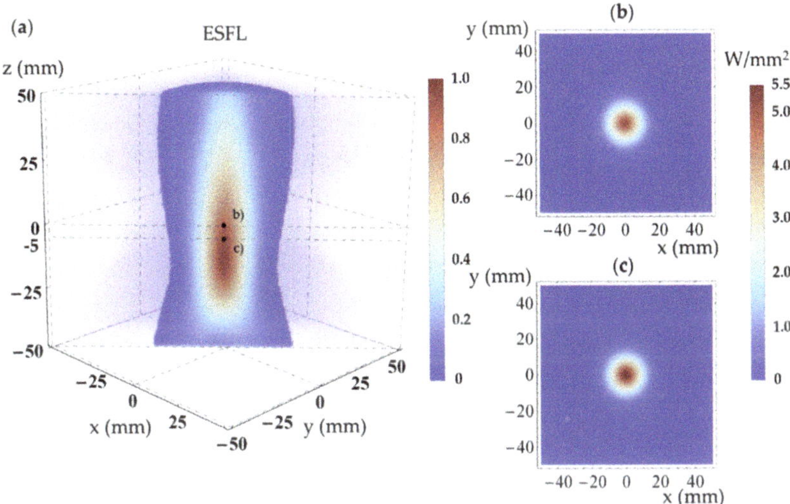

Figure 17. (**a**) Three-dimensional focal distribution of the ESFL. (**b**) Top view of the light distribution at the focal point (Z = 0 mm). (**c**) Top view of the light distribution with the highest concentrated solar flux (Z = −5 mm).

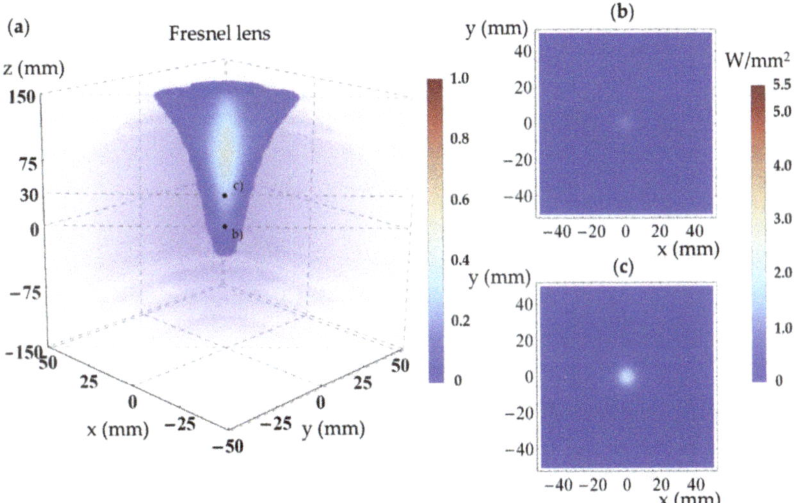

Figure 18. (**a**) Three-dimensional focal distribution of the flat Fresnel lens. (**b**) Top view of the light distribution at the focal point (Z = 0 mm). (**c**) Top view of the light distribution with the highest concentrated solar flux (Z = 30 mm).

4.2. Temperature Analysis of Both the ESFL and the Flat Fresnel Lens

Zemax non-sequential ray-tracing and Ansys finite element analyses were both used to evaluate the temperature of both the ESFL and the flat Fresnel lens at the Z position with the highest concentration solar flux found in Section 4.1. Ansys allows thermo-optical calculations that deal with complex geometric shape and boundary conditions, enabling the approximation of variables in a volume or surface element that changes across the matrix [46–48].

In Zemax, a square absorbing black body (emissivity $\varepsilon = 1$ [49]) detector of 20×20 mm^2 and 150×150 pixels was used to calculate the concentrated solar flux at the optimal focal positions of the ESFL and the flat Fresnel lens. The matrixial data were then exported to the Ansys workbench through the "External data" component and loaded as a heat flux source.

In Ansys 2021 finite element analysis, a graphite disk receiver of 20 mm diameter and 5 mm thickness was used to obtain the temperature of both concentrators. The graphite of 2250 kg/m^3 constant density, 24 W/m K thermal conductivity and 709 J/Kg K specific heat were chosen from the Ansys internal material library. The disk receiver was divided by the tetrahedrons meshing method with a sizing element of 0.4 mm. It contained approximately 1600 elements, which were enough for FEA calculations, with good approximation. The boundary conditions set for the convection applied onto the disk were the same as the natural stagnant air convection, representing a heat transfer coefficient of 5×10^{-6} W/mm^2/K. The radiation exchange between surfaces was restricted by a gray-diffused surface and the emissivity for the graphite disk surface was confined to $\varepsilon = 0.85$. A room temperature of 295.15 K was considered.

The respective temperatures of the ESFL and the flat Fresnel lens are shown in Figure 19. Both temperatures were generated from the focal distributions of Figures 17c and 18b, respectively. The ESFL attained maximum and minimum temperatures of 2362 K and 1945 K, respectively, which were 1.73 and 1.60 times more than that of the Fresnel lens with the maximum temperature of 1363 K and minimum temperature of 1217 K, respectively.

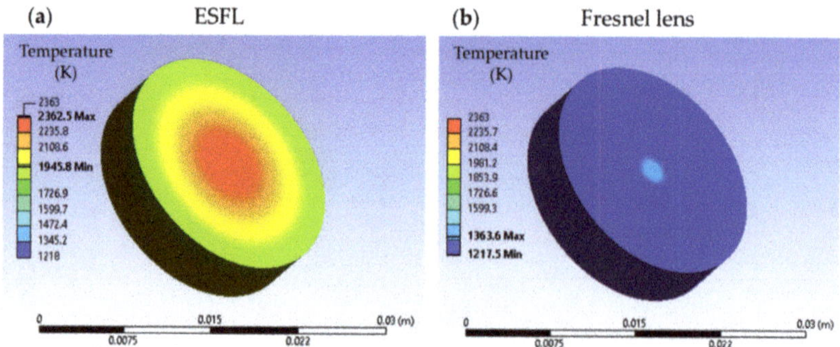

Figure 19. Focal temperature of ESFL (**a**) and flat Fresnel lens (**b**).

5. Conclusions

To overcome the aberration chromatic issue of Fresnel lens concentrators, a novel parametric model of a three-dimensional elliptical-shaped Fresnel lens was provided and analyzed. The modeling process took into account the solar Gaussian distribution based on the measured parameters by Vittitoe and Biggs [35], instead of the classical edge-ray principle method and the solar acceptance half-angle of 0.27°. The design was performed by CAD software and then imported into Zemax for numerical calculations. The accuracy of this model was confirmed by the numerical analysis of the output performance of a flat Fresnel, whose results matched well with the experimental data [36]. The ESFL output performance was compared to these results, under the same focusing conditions, revealing the advantage of the ESFL in focusing more efficiently the solar rays from the external annulus area, at the external annulus collection area, where a significant part of the incoming solar power is collected, resulting in a significant increase in the concentrated solar flux.

The study of different combinations of ESFL focal length and arch height, and their influence on the output performance was carried out. The highest peak solar flux of about 5.2 W/m^2 was attained for the ESFL with a shorter focal length of $h_f = 200$ mm and high arch height of $h_l = 500$ mm, resulting in a large aspect ratio of 0.603. As h_f increased, the solar concentration decreased, but maximum solar fluxes were attained with a smaller h_l, hence lowering the aspect ratio.

The optimal concentrated solar flux value within the focal cone of the best ESFL was then studied and compared to that of a flat Fresnel lens at its best output performance configuration. This study demonstrated the effectiveness of the ESFL in reducing the chromatic aberration, leading to a significant enhancement of the concentrated solar flux and temperature, as compared to that of the flat Fresnel lens.

In addition, the present work also provided a comprehensive study of the influence of the number of grooves and size on the ESFL performance. It was found that the optimal number of groves per millimeter could be substantially reduced in relation to that reported by the previous literature and market. This could greatly facilitate the manufacturing process of Fresnel lens concentrators while increasing its solar concentration capacity, revealing the promising potential of ESFL in many solar energy research and applications.

Author Contributions: Conceptualization, D.G.; methodology, D.G.; software, D.G., D.L., C.R.V., H.C. and M.C.; validation, D.G. and D.L.; formal analysis, D.G., D.L., C.R.V. and J.A.; investigation, D.G.; resources, D.L. and J.A.; data curation, D.G. and D.L.; writing—original draft preparation, D.G.; writing—review and editing, D.G., D.L., C.R.V., M.C., B.D.T., H.C. and J.A.; visualization, D.G.; supervision, D.L.; project administration, D.L.; funding acquisition, D.L. All authors have read and agreed to the published version of the manuscript.

Funding: This research was funded by Fundação para a Ciência e a Tecnologia—Ministério da Ciência, Tecnologia e Ensino Superior, grant number UIDB/00068/2020.

Institutional Review Board Statement: Not applicable.

Informed Consent Statement: Not applicable.

Data Availability Statement: Not applicable.

Acknowledgments: The contract CEECIND/03081/2017 and the fellowship grants SFRH/BD/145322 /2019, PD/BD/142827/2018, PD/BD/128267/2016 and SFRH/BPD/125116/2016 of Joana Almeida, Miguel Catela, Dário Garcia, Bruno D. Tibúrcio, and Cláudia R. Vistas, respectively, are acknowledged.

Conflicts of Interest: The authors declare no conflict of interest. The funders had no role in the design of the study; in the collection, analyses, or interpretation of data; in the writing of the manuscript, or in the decision to publish the results.

References

1. Price, J.S.; Sheng, X.; Meulblok, B.M.; Rogers, J.A.; Giebink, N.C. Wide-angle planar microtracking for quasi-static microcell concentrating photovoltaics. *Nat. Commun.* **2015**, *6*, 6223. [CrossRef] [PubMed]
2. Coughenour, B.M.; Stalcup, T.; Wheelwright, B.; Geary, A.; Hammer, K.; Angel, R. Dish-based high concentration PV system with Köhler optics. *Opt. Express* **2014**, *22*, A211–A224. [CrossRef] [PubMed]
3. El Majid, B.; Motahhir, S.; El Ghzizal, A. Parabolic bifacial solar panel with the cooling system: Concept and challenges. *SN Appl. Sci.* **2019**, *1*, 1176. [CrossRef]
4. Xie, W.T.; Dai, Y.J.; Wang, R.Z.; Sumathy, K. Concentrated solar energy applications using Fresnel lenses: A review. *Renew. Sustain. Energy Rev.* **2011**, *15*, 2588–2606. [CrossRef]
5. Cosby, R.M. *Solar Concentration by Curved-Base Fresnel Lenses*; Ball State University: Muncie, IN, USA, 1977.
6. Neill, M.J.O.; McDanal, A.J. Manufacturing Technology Improvements for a Line-Focus Concentrator Module. In Proceedings of the Twenty Third IEEE Photovoltaic Specialists Conference—1993, Conference Record (Cat. No.93CH3283-9). Louisville, KY, USA, 10–14 May 1993; pp. 1082–1089.
7. Kritchman, E.M.; Friesem, A.A.; Yekutieli, G. Highly Concentrating Fresnel Lenses. *Appl. Opt.* **1979**, *18*, 2688–2695. [CrossRef]
8. Leutz, R.; Suzuki, A.; Akisawa, A.; Kashiwagi, T. Design of a nonimaging Fresnel lens for solar concentrators. *Sol. Energy* **1999**, *65*, 379–387. [CrossRef]
9. Yeh, N. Analysis of spectrum distribution and optical losses under Fresnel lenses. *Renew. Sustain. Energy Rev.* **2010**, *14*, 2926–2935. [CrossRef]
10. Akisawa, A.; Hiramatsu, M.; Ozaki, K. Design of dome-shaped non-imaging Fresnel lenses taking chromatic aberration into account. *Sol. Energy* **2012**, *86*, 877–885. [CrossRef]
11. Romero, M.; Steinfeld, A. Concentrating solar thermal power and thermochemical fuels. *Energy Environ. Sci.* **2012**, *5*, 9234–9245. [CrossRef]

12. Languy, F.; Habraken, S. Nonimaging achromatic shaped Fresnel lenses for ultrahigh solar concentration. *Opt. Lett.* **2013**, *38*, 1730–1732. [CrossRef]
13. Ma, X.; Zheng, H.; Tian, M. Optimize the shape of curved-Fresnel lens to maximize its transmittance. *Sol. Energy* **2016**, *127*, 285–293. [CrossRef]
14. Leutz, R.; Suzuki, A. *Nonimaging Fresnel Lenses: Design and Performance of Solar Concentrators*; Springer Science & Business Media: Berlin/Heidelberg, Germany, 2001.
15. Kalogirou, S. *Solar Energy Engineering Processes and Systems*; Academic Press: Cambridge, MA, USA, 2013.
16. Rabl, A. *Active Solar Collectors and Their Applications*; Oxford University Press on Demand: Oxford, UK, 1985.
17. Winston, R. Principles of Solar Concentrators of a Novel Design. *Sol. Energy* **1974**, *16*, 89–95. [CrossRef]
18. Cheng, Y.; Zhang, X.D.; Zhang, G.X. Design and machining of Fresnel solar concentrator surfaces. *Int. J. Precis. Technol.* **2013**, *3*, 354–369. [CrossRef]
19. Zheng, H.; Feng, C.; Su, Y.; Dai, J.; Ma, X. Design and experimental analysis of a cylindrical compound Fresnel solar concentrator. *Sol. Energy* **2014**, *107*, 26–37. [CrossRef]
20. Viera-González, P.M.; Sánchez-Guerrero, G.E.; Martínez-Guerra, E.; Ceballos-Herrera, D.E. Mathematical Analysis of Nonimaging Fresnel Lenses Using Refractive and Total Internal Reflection Prisms for Sunlight Concentration. *Math. Probl. Eng.* **2018**, *2018*, 4654795. [CrossRef]
21. Yeh, N. Illumination uniformity issue explored via two-stage solar concentrator system based on Fresnel lens and compound flat concentrator. *Energy* **2016**, *95*, 542–549. [CrossRef]
22. Yeh, N.; Yeh, P. Analysis of point-focused, non-imaging Fresnel lenses' concentration profile and manufacture parameters. *Renew. Energy* **2016**, *85*, 514–523. [CrossRef]
23. Ma, X.; Jin, R.; Liang, S.; Zheng, H. Ideal shape of Fresnel lens for visible solar light concentration. *Opt. Express* **2020**, *28*, 18141–18149. [CrossRef]
24. Yeh, N. Optical geometry approach for elliptical Fresnel lens design and chromatic aberration. *Sol. Energy Mater. Sol. Cells* **2009**, *93*, 1309–1317. [CrossRef]
25. Fresneltech Brochure. Available online: https://www.fresneltech.com/fresnel-lenses (accessed on 17 June 2021).
26. Garcia-Segura, A.; Fernandez-Garcia, A.; Ariza, M.J.; Sutter, F.; Valenzuela, L. Durability studies of solar reflectors: A review. *Renew. Sustain. Energy Rev.* **2016**, *62*, 453–467. [CrossRef]
27. He, C.; Duan, X.; Zhao, Y.; Feng, J. An analytical flux density distribution model with a closed-form expression for a flat heliostat. *Appl. Energy* **2019**, *251*, 113310. [CrossRef]
28. Garcia, D.; Liang, D.; Tibúrcio, B.D.; Almeida, J.; Vistas, C.R. A three-dimensional ring-array concentrator solar furnace. *Sol. Energy* **2019**, *193*, 915–928. [CrossRef]
29. Lv, H.; Huang, X.; Li, J.; Huang, W.; Li, Y.; Su, Y. Non-uniform sizing of PV cells in the dense-array module to match the non-uniform illumination in dish-type CPV systems. *Int. J. Low-Carbon Technol.* **2020**, *15*, 565–573. [CrossRef]
30. Garcia, D.; Liang, D.; Almeida, J.; Tibúrcio, B.D.; Costa, H.; Catela, M.; Vistas, C.R. Analytical and numerical analysis of a ring-array concentrator. *Int. J. Energy Res.* **2021**, *45*, 15110–15123. [CrossRef]
31. Yeh, P.; Yeh, N. Design and analysis of solar-tracking 2D Fresnel lens-based two staged, spectrum-splitting solar concentrators. *Renew. Energy* **2018**, *120*, 1–13. [CrossRef]
32. Pan, J.-W.; Huang, J.-Y.; Wang, C.-M.; Hong, H.-F.; Liang, Y.-P. High concentration and homogenized Fresnel lens without secondary optics element. *Opt. Commun.* **2011**, *284*, 4283–4288. [CrossRef]
33. Zhao, Y.; Zheng, H.; Sun, B.; Li, C.; Wu, Y. Development and performance studies of a novel portable solar cooker using a curved Fresnel lens concentrator. *Sol. Energy* **2018**, *174*, 263–272. [CrossRef]
34. Shen, F.; Huang, W. Study on the Optical Properties of the Point-Focus Fresnel System. *Sustainability* **2021**, *13*, 10367. [CrossRef]
35. Vittitoe, C.N.; Biggs, F. Six-gaussian representation of the angular-brightness distribution for solar radiation. *Sol. Energy* **1981**, *27*, 469–490. [CrossRef]
36. Ferriere, A.; Rogriguez, G.; Sobrino, J. Flux Distribution Delivered by a Fresnel Lens Used for Concentrating Solar Energy. *J. Sol. Energy Eng.* **2004**, *126*, 654–660. [CrossRef]
37. ASTM International Standard Tables for Reference Solar Spectral Irradiance at Air Mass 1.5: Direct Normal and Hemispherical for a 37 Degree Tilted Surface. Available online: https://www.nrel.gov/grid/solar-resource/spectra-am1.5.html (accessed on 21 December 2021).
38. MEÉSU Procédés. MSSF horizontal—PROMES. Available online: https://www.promes.cnrs.fr/index.php?page=mssf-horizontal (accessed on 7 March 2019).
39. Flamant, G.; Ferriere, A.; Laplaze, D.; Monty, C. Solar Processing of Materials: Opportunities and New Frontiers. *Sol. Energy* **1999**, *66*, 117–132. [CrossRef]
40. Ferriere, A.; Sanchez Bautista, C.; Rodriguez, G.P.; Vazquez, A.J. Corrosion resistance of stainless steel coatings elaborated by solar cladding process. *Sol. Energy* **2006**, *80*, 1338–1343. [CrossRef]
41. Gineste, J.M.; Flamant, G.; Olalde, G. Incident solar radiation data at Odeillo solar furnaces. *J. Phys. IV* **1999**, *9*, Pr3-623–Pr3-628. [CrossRef]
42. Liang, K.; Zhang, H.; Chen, H.; Gao, D.; Liu, Y. Design and test of an annular fresnel solar concentrator to obtain a high-concentration solar energy flux. *Energy* **2021**, *214*, 118947. [CrossRef]

43. Leutz, R.; Suzuki, A.; Akisawa, A.; Kashiwagi, T. Shaped nonimaging Fresnel lenses. *J. Opt. A Pure Appl. Opt.* **2000**, *2*, 112–116. [CrossRef]
44. Nelson, D.T.; Evans, D.L.; Bansal, R.K. Linear Fresnel Lens Concentrators. *Sol. Energy* **1975**, *17*, 285–289. [CrossRef]
45. Sierra, C.; Vázquez, A. NiAl coatings on carbon steel by self-propagating high-temperature synthesis assisted with concentrated solar energy: Mass influence on adherence and porosity. *Sol. Energy Mater. Sol. Cells* **2005**, *86*, 33–42. [CrossRef]
46. An, W.; Ruan, L.M.; Qi, H.; Liu, L.H. Finite element method for radiative heat transfer in absorbing and anisotropic scattering media. *J. Quant. Spectrosc. Radiat. Transf.* **2005**, *96*, 409–422. [CrossRef]
47. Li, B.; Oliveira, F.A.C.; Rodriguez, J.; Fernandes, J.C.; Rosa, L.G. Numerical and experimental study on improving temperature uniformity of solar furnaces for materials processing. *Sol. Energy* **2015**, *115*, 95–108. [CrossRef]
48. Rinker, G.; Solomon, L.; Qiu, S.G. Optimal placement of radiation shields in the displacer of a Stirling engine. *Appl. Therm. Eng.* **2018**, *144*, 65–70. [CrossRef]
49. Radiant Zemax. *Zemax Manual*; Radiant Zemax: Sacramento, CA, USA, 2014; Volume 13.

Article

Highly Efficient Solar Laser Pumping Using a Solar Concentrator Combining a Fresnel Lens and Modified Parabolic Mirror

Zitao Cai, Changming Zhao *, Ziyin Zhao, Xingyu Yao, Haiyang Zhang and Zilong Zhang

School of Optics and Photonics, Beijing Institute of Technology, No. 5, South Street, Zhongguancun, Beijing 100081, China; 3120185330@bit.edu.cn (Z.C.); 3220200493@bit.edu.cn (Z.Z.); 3220190441@bit.edu.cn (X.Y.); ocean@bit.edu.cn (H.Z.); zlzhang@bit.edu.cn (Z.Z.)
* Correspondence: zhaochangming@bit.edu.cn

Abstract: Solar-pumped lasers (SPLs) allow direct solar-to-laser power conversion, and hence, provide an opportunity to harness a renewable energy source. Herein, we report significant improvements in end-side-pumped solar laser collection efficiency and beam brightness using a novel 1.5-m-diameter compound solar concentrator combining a Fresnel lens and modified parabolic mirror. A key component of this scheme is the off-axis-focused parabolic mirror. An original dual-parabolic pump cavity is another feature. To determine the dependence of the SPL performance on the distance between the focus and central axis of the modified parabolic mirror, several systems with different distances were optimized using TracePro and ASLD software. It was numerically calculated that end-side pumping a 5-mm-diameter, 22-mm-long Nd:YAG crystal rod would generate 74.6 W of continuous-wave solar laser power at a collection efficiency of 42.2 W/m^2, i.e., 1.1 times greater than the previous record value. Considering the laser beam quality, a brightness figure of 0.063 W was obtained, which is higher than that of other multimode SPL designs with end-side pumping. Thus, our SPL concentrator offers the possibility of achieving a beam quality as high as that obtainable via side pumping, alongside highly efficient energy conversion, which is characteristic to end-side pumping.

Keywords: solar pumping; laser; parabolic mirror; Fresnel lens; solar concentrator; Nd:YAG

Citation: Cai, Z.; Zhao, C.; Zhao, Z.; Yao, X.; Zhang, H.; Zhang, Z. Highly Efficient Solar Laser Pumping Using a Solar Concentrator Combining a Fresnel Lens and Modified Parabolic Mirror. *Energies* **2022**, *15*, 1792. https://doi.org/10.3390/en15051792

Academic Editor: Tapas Mallick

Received: 5 February 2022
Accepted: 25 February 2022
Published: 28 February 2022

Publisher's Note: MDPI stays neutral with regard to jurisdictional claims in published maps and institutional affiliations.

1. Introduction

Solar energy, as a desirable renewable energy source, has become increasingly important because of the need to reduce fossil fuel consumption and the deterioration of the global environment. Solar-pumped lasers (SPLs) can directly convert light from the sun into laser energy without the requirement for electrically driven artificial pump light sources. The use of these lasers facilitates greater solar energy utilization and has attracted the attention of researchers in the fields of renewable energy and laser technology. Many potential opportunities for SPL use in solar hydrogen generation [1–3] and space laser [4–6] applications have been predicted.

Because of their applicability, numerous researchers have conducted studies on SPLs. Young reported the first continuous-wave Nd:YAG SPL emission with a power of 1 W in 1966 [7]. To obtain appreciable laser output power, parabolic mirrors with large collection areas, for example, 78.5 m^2 [8], 50.2 m^2 [9], 38.5 m^2 [10], and 6.75 m^2 [11], were used to concentrate solar radiation on laser media in early studies. These primary concentrators with parabolic mirrors were very large and expensive, and the mechanical supporting structures of the laser cavities inside the parabolic mirrors produced shadows that reduced the effective solar energy collection area. In the last decade, the adoption of Fresnel lens designs has accelerated the development of low-cost solar lasers with reduced weight. As a consequence, many studies focusing on obtaining higher laser collection efficiency, which is defined as the laser output power per unit area of the solar collector, have been conducted in recent years. In 2007, Yabe et al. obtained an 18.7 W/m^2 laser collection efficiency by end pumping a 9-mm-diameter, 100-mm-long Cr:Nd:YAG ceramic rod using a 1.3-m^2 Fresnel lens [12]. In 2011, a laser collection efficiency of 19.3 W/m^2 was achieved by Liang and Almeida using a 0.64-m^2 Fresnel lens to end-side pump a 4-mm-diameter, 25-mm-long Nd:YAG crystal rod [13]. In 2012, Dinh et al. reported a 30-W/m^2 laser collection efficiency when end-side pumping a 6-mm-diameter, 100-mm-long Nd:YAG crystal using a 4-m^2 Fresnel lens [14]. More recently, in 2018, Guan et al. achieved a 32.1-W/m^2 laser collection efficiency with the use of a 1.03-m^2 Fresnel lens for end-side pumping a 6-mm-diameter, 95-mm-long Nd:YAG–YAG-bonded crystal rod [15]. However, because of the defects of Fresnel lenses such as dispersion, which limits the geometric concentration ratio, Fresnel lenses have not completely replaced parabolic mirrors in SPL designs. For example, in 2018, Liang et al. reported a laser collection efficiency of 32.5 W/m^2 when end pumping a 4.5-mm-diameter, 35-mm-long Cr:Nd:YAG ceramic rod using a heliostat-parabolic mirror with a 1.0-m^2 effective collection area, which represents the record for highest solar laser collection efficiency [16]. In addition, sunlight is more focused when concentrated by parabolic mirrors rather than Fresnel lenses, because of the absence of dispersion. Thus, compared with SPLs with Fresnel lenses, SPLs with parabolic mirrors typically have a better brightness figure of merit [17,18], which is defined as the ratio between laser power and the product of the M_x^2 and M_y^2 factors [11]. The use of a parabolic mirror in an SPL design also makes it compatible with side pumping to produce a TEM$_{00}$-mode output [19,20].

Either a Fresnel lens or a parabolic mirror is used as the primary concentrator in almost all SPLs. However, as mentioned above, a drawback of the former is dispersion, whereas when the latter is used, the laser cavity shadows the incoming solar light. Hence, neither option is the perfect choice for the primary concentrator of a SPL. Therefore, it is very important to develop new solar-energy collection and concentration systems. In recent years, many novel solar concentrators have been proposed, and numerically calculated laser performances for their use in SPLs have been reported [21–25]. Among these, a modified ring-array concentrator (RAC) allowed some remarkable laser performances to be achieved. When a small Fresnel lens was added to the center of the RAC, laser collection efficiencies of 38.4 and 29.18 W/m^2 were obtained via end-side pumping and side pumping a Nd:YAG crystal, respectively [22,25].

A novel 1.5-m-diameter solar concentrator composed of a modified parabolic mirror and Fresnel lens mounted coaxially is proposed in this paper. To avoid shadowing in the laser cavity, a section of the parabolic mirror behind the focal point is removed. Thus, the total solar collection area, including the remaining annular parabolic mirror section and the Fresnel lens, spans 1.766 m^2. As a compound structure, this solar concentrator combines the advantages of Fresnel lenses and parabolic mirrors. When compared with a parabolic mirror concentrator, the replaced part with a Fresnel lens produces a solar concentrator that is less heavy and shadowing does not occur. When compared with a typical Fresnel lens concentrator, the reduced diameter of the Fresnel lens in this compound system means that the influence of dispersion is reduced. Further, because the Fresnel lens and parabolic mirror pump the laser medium from the end and side, respectively, flexible adjustment of the solar power distribution in the laser medium is possible. It should also be mentioned that because the focus of the modified parabolic mirror is off-axis, serious heat accumulation in the laser medium is prevented.

A series of optimizations performed using the TracePro and ASLD simulation software packages ensured that, when combined with a novel dual-parabolic pump cavity, the solar radiation collected by this solar concentrator was uniformly and efficiently absorbed by a 5-mm-diameter, 22-mm-long Nd:YAG single-crystal laser rod. The continuous-wave solar laser output power was 74.6 W, and laser beam quality factors of M_x^2 = 34.0 and M_y^2 = 35.1 were obtained numerically; the corresponding laser collection efficiency was 42.2 W/m^2, and the beam brightness figure of merit was 0.063 W, i.e., 1.1 and 2.6 times greater than the previous records, respectively, for an end-side pumped SPL [22]. The structure model, design principles, and optimization of the detailed parameters are explained in

Section 2. The laser performance, numerically calculated by ASLD software, is explained in Section 3. A discussion of the results is presented in Section 4.

2. Method

The solar laser output power (P_{out}) can be analytically calculated in terms of absorbed solar power (P_{ab}) and other quantities by Equation (1) [26],

$$P_{\text{out}} = \frac{1-R}{1+R} A I_S \left[\frac{2\eta_Q \eta_S \eta_B P_{ab}}{(2\alpha l + L_M - \ln R) A I_S} - 1 \right] \tag{1}$$

which is considered as the overall theoretical design method. Here, R is the output coupler reflectivity, A is the cross-sectional area of the laser rod, I_S is the saturation intensity, and η_Q, η_S, and η_B are the quantum efficiency, Stokes' factor, and beam overlap efficiency, respectively. The factor $2\alpha l$ accounts for the two-way absorption and scattering loss of the laser medium inside the laser resonator, that is, α is the scattering and absorption coefficient of the laser material, and L_M represents other losses in the resonator. It is evident that the solar laser output power (P_{out}) has a positive correlation with the absorbed solar power (P_{ab}), and a negative correlation with the length (l) of the laser medium. In addition, the thermal effects caused by inhomogeneous solar absorption distributions also affect the solar laser output power. Hence, this approach is aimed to achieve a homogeneous distribution and efficient solar radiation absorption in the laser rod with a shorter length.

The model of the SPL proposed in this paper (Figure 1) is composed of two major parts, the solar energy collection and concentration system and the solar laser head. The structure of these two parts and their optical principles are explained in Sections 2.1 and 2.2, respectively. On this basis, more detailed parameters in this approach are optimized in TracePro to obtain an efficient solar radiation absorption, and this process is presented step by step in Section 2.3.

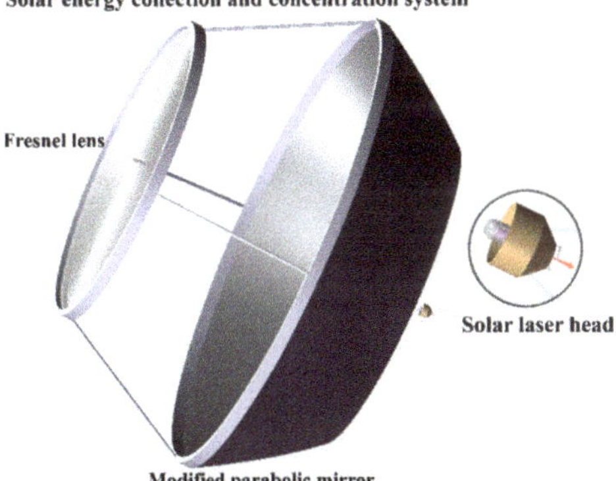

Figure 1. Model of the solar-pumped laser consisting of the solar energy collection and concentration system and the solar laser head.

2.1. Solar Energy Collection and Concentrator System

The structure of the primary concentrator proposed in this paper (Figure 2) consists of a Fresnel lens and a modified parabolic mirror with an aperture of 1.5 m. The Fresnel lens was placed at the top of the concentrator system to concentrate the sunlight at the front end of the laser head; the diameter and focal length were both 1 m. The parabolic mirror was mounted coaxially below the Fresnel lens to concentrate sunlight onto the sides of the laser head; its parabolic surface was silver-coated (reflectivity $R = 95\%$ for sunlight). The inner and outer diameters of the modified parabolic mirror were 1.5 and 1 m, respectively, and the other parameters were numerically optimized using TracePro optical simulations based on geometric optics analysis.

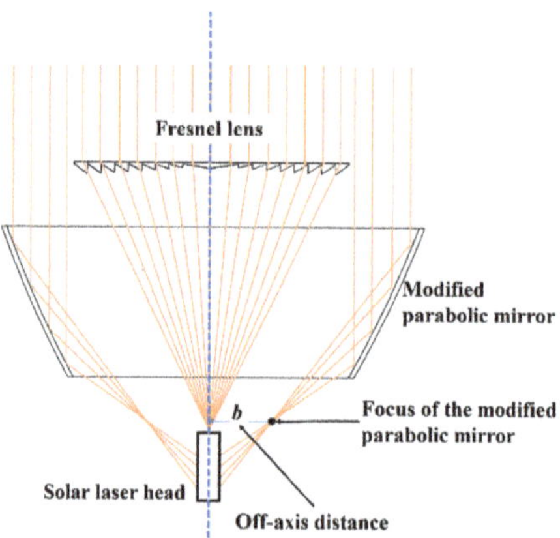

Figure 2. Schematic of the primary concentrator consisting of a Fresnel lens and modified parabolic mirror.

The schematic diagrams in Figure 3 show a conventional parabolic mirror (Figure 3a) for comparison alongside the proposed modified parabolic mirror, for which the focus is off-axis (Figure 3b,c). In general, the reflective surface of a parabolic mirror takes the form of a parabola with the equation $x^2 = 2py$ in the XOY plane that is rotated around the Y axis (Figure 3a), and this structure concentrates light at the focal point $(p/2,0)$. However, when the collection area of the parabolic mirror is large and most of the optical power is highly concentrated at the focal position, serious heat accumulation in the laser medium can occur. Therefore, we moved the parabola horizontally outward and inward, varying the distance b between the focus and central axis (Figure 3b,c). This design means that along the central axis, the light is slightly unfocused, being longitudinally stretched; thus, the average solar power in the concentrated area is reduced, rendering it suitable for side pumping the laser medium.

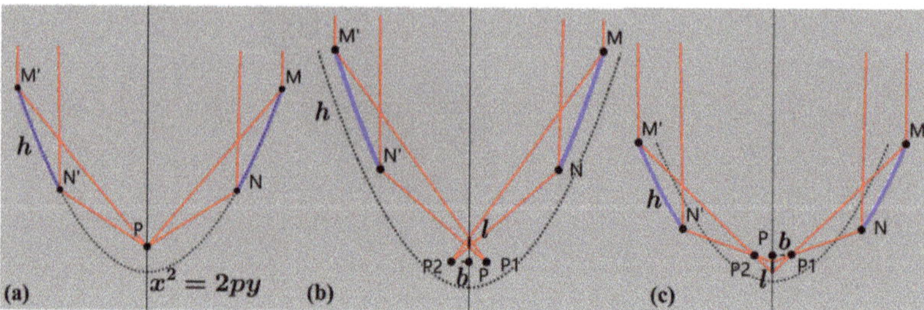

Figure 3. Parabolic mirror schematics: (**a**) Principle of conventional parabolic mirror; (**b**) Parabolic mirror with inwardly deviated focus; (**c**) Parabolic mirror with outwardly deviated focus.

Laser media end pumped by sunlight from primary concentrators in SPLs always have inhomogeneous solar absorption distributions, with strong absorption at the top that is gradually reduced along the axial direction of the laser media [15,16,18]. The Fresnel lens in the primary concentrator is expected to produce this type of distribution in the laser medium. When sunlight is incident parallel to the central axis on the parabolic mirror, the mirror surface near the upper edge reflects the sunlight at a smaller convergence angle. Therefore, when the parabola is moved inward (Figure 3b), an absorber on the central axis sits in front of the focus, and its solar absorption distribution weakens gradually from top to bottom along the axial direction (Figure 4a). Conversely, when the parabola is moved outward (Figure 3c), the focus is in front of the absorber, and the opposite distribution occurs (Figure 4b); in this case, a laser medium on the central axis absorbs little power at the top, but this gradually increases along the axial direction toward the bottom. Thus, the parabolic mirror should be moved outward to concentrate sunlight on the side of the laser medium and produce a solar absorption distribution opposite to that of the Fresnel lens. Owing to the complementarity of the Fresnel lens and modified parabolic mirror, the compound primary concentrator can generate a uniform solar absorption distribution in the laser medium, and the reduction in heat accumulation in the laser medium means that a laser output with better beam quality factors can be expected.

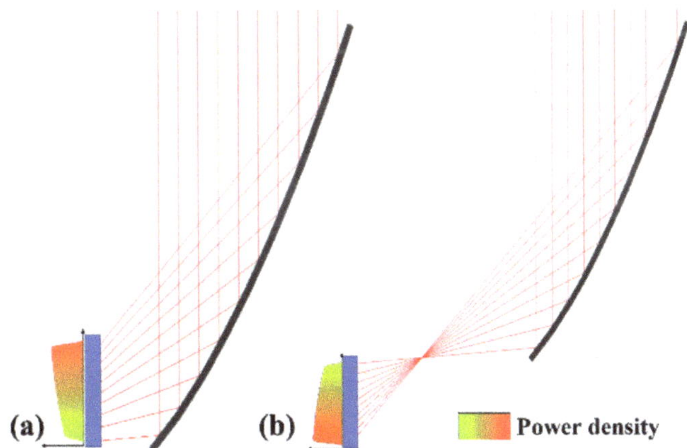

Figure 4. Solar absorption distributions of absorbers along the axial direction: (**a**) A absorber in front of the focus; (**b**) A absorber behind the focus.

The focal length $p/2$ and distance between the focus and central axis b, which affects the incident angle and position of the solar light concentrated on the laser head, were optimized using TracePro. After outward deviation from the central axis, the curve equation of the parabolic mirror is $(x - b)^2 = 2py$ and the focus occurs at $(b, p/2)$, on the right side of the XOY plane (Figure 3c). To avoid shadowing, $p/2$ should take an appropriate value such that the laser cavity can be placed behind the focal point. At the lower boundary of the parabolic mirror, the focal length should satisfy the condition $(500 - b)^2/2p \geq p/2$, i.e., $p \leq 500 - b$. When the value of p is close to the maximum value, the longitudinal length of the parabolic mirror is small; thus, the weight of the system is reduced. Moreover, in this case, the incident angles of the sunlight concentrated by the parabolic mirror on the laser medium will be smaller, which will reduce the transmission loss of pump light on the surface of the laser medium.

2.2. Solar Laser Head

The solar laser head of this system was composed of a dual-parabolic-surface pump cavity and a liquid light guide lens (LLGL) (Figure 5), within which was mounted a 5-mm-diameter Nd:YAG single-crystal laser rod with a high-reflection (HR) coating optimized for 1064 nm on the front face and an antireflection (AR) coating for the same wavelength on the back face. The output coupler behind the laser rod was coated for partial reflection (PR) at 1064 nm, and the laser resonator was bounded by a PR output coupler and HR-coated mirror.

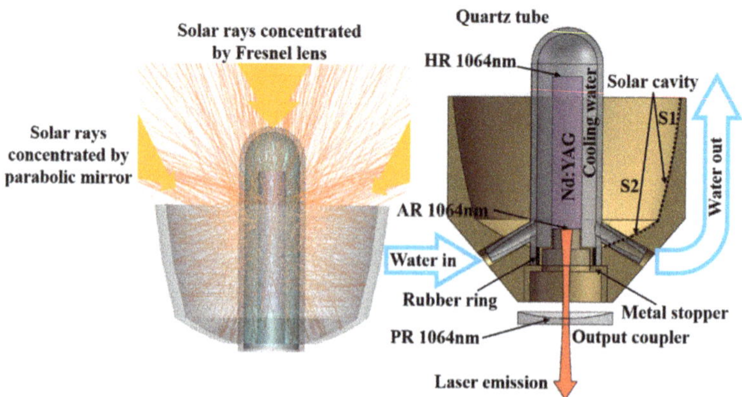

Figure 5. Liquid light-guide lens (LLGL) and dual-parabolic pump cavity containing 5-mm-diameter Nd:YAG rod. The incident solar light, 1064-nm laser resonator, water cooling scheme, and locations of the two parabolic surfaces in the solar cavity are illustrated.

The inner wall of the pump cavity was silver-coated to achieve 95% sunlight reflectivity. Homogeneous distribution and efficient solar radiation absorption in the laser rod were generated owing to the fact that the inner wall of the pump cavity consisted of two parabolic surfaces S1 and S2 (Figure 5). Sunlight that had been concentrated by the parabolic mirror but transmitted through the laser medium without being fully absorbed was mostly incident on the side surface of the pump cavity. The parabolic surface S1 had the form of a left-opening parabola with the curve equation in the XOY plane $(y - p/2)^2 = -2m(x - (m/2 - b))$ that is rotated around the Y axis (Figure 6). Because S1 was focused at the same point as the parabolic mirror of the primary concentrator, from the opposite direction, sunlight beams concentrated by the parabolic mirror (red lines in Figure 6) passing through the laser rod, were reflected in parallel by the paraboloid mn (yellow lines in Figure 6), and then passed through the laser rod for a second time; these beams were finally reflected by the paraboloid m′n′ on the other side of the central axis (green lines in Figure 6), and were incident on the laser rod for a third time. This parabolic cavity provides an isosceles triangle pathway for sunlight, which allows a short length of laser rod to fully absorb the incident sunlight. Solar light that had been concentrated by the Fresnel lens and passed through the laser medium without being fully absorbed was mostly incident on the end surface of the pump cavity, which took the form of an upward-opening parabolic surface (S2). All the parameters of S1 and S2 were optimized using TracePro.

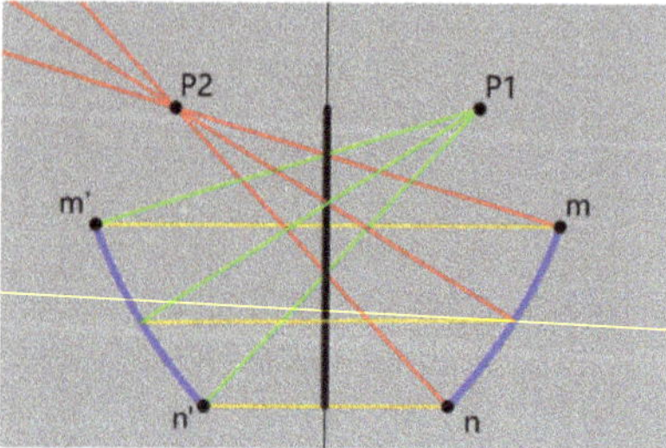

Figure 6. Schematic of the parabolic surface S1 showing the isosceles-triangle pathway for sunlight using different colors. Red, yellow, and green lines represent the solar rays before reflection in the cavity, after the first reflection, and after the second reflection, respectively.

The LLGL consisted of a quartz tube filled with water for cooling. When filled with water, the head and side of the LLGL act as spherical and cylindrical lenses, respectively, enlarging the effective absorption areas of the end and side surfaces of the laser media. The water was pumped by an external water pump to cool the laser medium. The inner wall of the quartz tube had a diameter of 10 mm, and the space around the laser rod was deemed sufficient for efficient water cooling of both the laser medium and the LLGL. The bottom opening of the quartz tube was blocked by an internal hollow metal stopper, which was connected to the end of the laser rod by a 5-mm-diameter hollow cylinder. The diameter of the inner hole was approximately 5 mm, and hence the laser resonator was hardly affected. The metal structure was partially silver-coated inside the quartz tube ($R = 95\%$ for sunlight), which allowed the laser rod length to be reduced such that the rod was shorter than the quartz tube and filled the void of the pump cavity, reflecting some of the sunlight back into the pump cavity.

2.3. Optimization of the Optical Design Parameters Using TracePro

To achieve the maximum absorbed solar pump power with the shortest laser rod length, the design parameters of the primary concentrator and laser head were optimized using the optical simulation software TracePro. For this system, several parameters were preset in TracePro, according to the actual experimental conditions. The Fresnel lens material was polymethyl methacrylate (PMMA), its diameter and focal length were both 1 m, the ring distance was 0.33 mm, and its thickness was 3 mm. The inner and outer diameters of the parabolic mirror section were 1.5 and 1 m, respectively, and the surface reflectivity was assumed to be 95%. The material of the quartz tube of the LLGL was fused silica, and the inner and outer diameters were set to 10 and 12 mm, respectively. A 5-mm-diameter Nd:YAG single-crystal laser rod was mounted coaxially in the quartz tube 8-mm away from the top of the spherical surface, and its initial length was set to 40 mm. A total of 50,000 rays distributed over a half-angle of $\pm0.27°$ were set within the 1.5-m aperture range to simulate incident sunlight, and the standard solar spectrum (air mass, AM1.5) was used in the calculations to distribute 980 W/m^2 of solar irradiance over the wavelengths of the solar spectrum (ASTM G173-03 International standard) in TracePro [27]. The absorption and dispersion spectra of all the materials were also defined, and the absorption spectrum of the Nd: YAG crystal was set assuming a Nd^{3+} doping concentration of 1%. Subsequently, the absorbed solar power was obtained by calculating the absorption and loss in the laser rod. Four steps for optimization of the modified parabolic mirror, parabolic surface S1 in the pump cavity, laser rod length, and parabolic surface S2 in the pump cavity were sequentially considered. According to the absorbed solar energy from the flux table generated by TracePro, these steps are outlined as follows.

2.3.1. Step 1: Optimization of the Modified Parabolic Mirror

When the offset distance b between the focus and the central axis is large, a large length of laser rod is needed to absorb a sufficient amount of sunlight, and when b is small, the sunlight is too concentrated and may damage the laser rod. Through a comparative analysis, it was determined that the optimal value of b was in the range of 20–50 mm. Therefore, four typical values for the offset distances, 20, 30, 40, and 50 mm, were used in the subsequent analysis; for comparison, a conventional parabolic mirror was also simulated. To distinguish the concentrator systems, they were labeled systems 1–5 according to the value of b.

Based on the analysis in Section 2.1, when the value of b is in the range of 20–50 mm, the focal length $p/2$ of the paraboloid has a maximum value of 225 mm for ideal parallel light incidence. However, because of the solar divergence angle of 0.27°, the focal length should be smaller, and after gradually reducing the value of $p/2$ in TracePro, it was found that a value of 200 mm was appropriate. The parameters of these parabolic mirrors and the solar power absorbed by the laser rods for the different systems (1–5) are shown in Table 1.

Table 1. Parameters of the modified parabolic mirror and absorbed solar power concentrated by the different primary concentrator systems and LLGL without considering the pump cavity.

System No.	1	2	3	4	5
Focal length $p/2$ (mm)	200	200	200	200	200
Deviation distance b (mm)	0	20	30	40	50
Distance above the laser head (mm)	109.5	89	78	68	58
Absorbed solar power (W)	177.2	157.1	157.2	157.2	157.2

2.3.2. Step 2: Optimization of Parabolic Surface S1 in the Pump Cavity

Ray tracing and visual analysis using TracePro allowed the positions and diameters of the input windows of the pump cavities to be rapidly determined. The surface equations of S1 can also be calculated using the expression defined in Section 2.2, $(y - p/2)^2 = -2m(x - (m/2 - b))$. As shown in Figure 7, there are clear boundaries defining the ranges over which the parabolic surface S1 receives sunlight, and this data can be used as the basis for determining the depth of S1 in the pump cavity. The position and diameter of the input windows of the pump cavities, the parameters of the parabolic surfaces S1, and the absorbed solar power in the laser rods when this structure is used with each of the different concentrators are listed in Table 2.

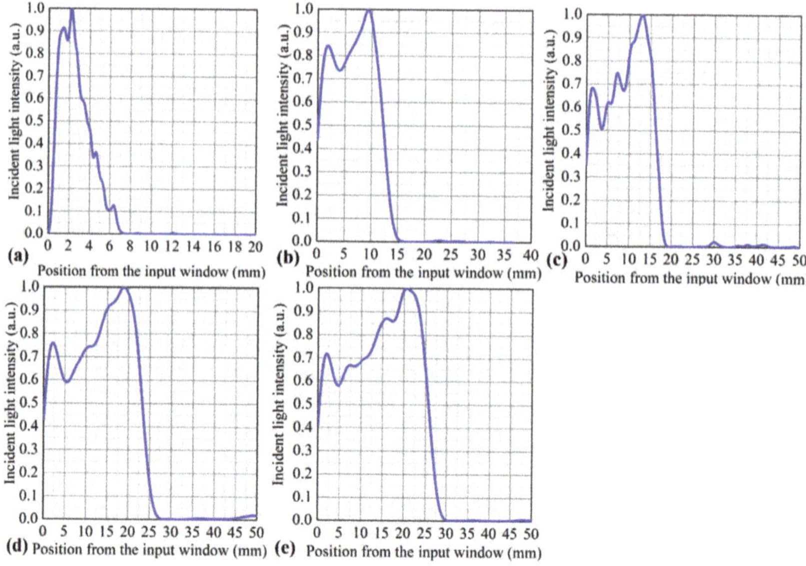

Figure 7. Distributions of incident solar light intensity on the surface of the pump cavity S1 for different systems: (**a**) System 1; (**b**) System 2; (**c**) System 3; (**d**) System 4; (**e**) System 5.

Table 2. Parameters of the parabolic surface S1 and absorbed solar power for each system investigated.

System No.	1	2	3	4	5
Distance behind the laser head (mm)	4.5	5	6	7	10
Diameter of input window (mm)	21.8	35.5	39.4	51.2	56.6
Focal length of the S1 parabola $m/2$ (mm)	11	38	50	66	79
Depth (mm)	7	15	18	26	30
Absorbed solar power (W)	241.2	229.8	226.0	222.2	219.7

2.3.3. Step 3: Optimization of the Laser Rod Length

As shown in Figure 8, when the length of the laser rod is 40 mm, parts of the rod absorb little solar power. Table 3 shows the laser rod length optimization results. Although the absorbed solar power in the laser rods decreased slightly as the laser rod length was increased, the absorption and scattering losses in the resonator were greatly reduced, which would result in increased solar-to-laser power conversion efficiencies.

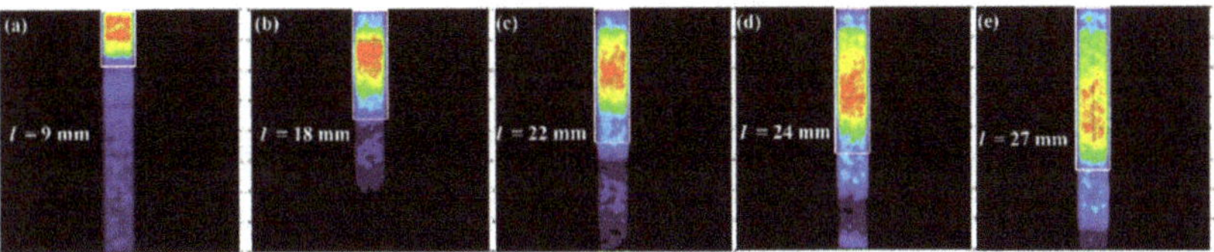

Figure 8. Incident solar power distributions along the axial direction of laser rods for the different systems (red area indicates the higher incident light intensity and the purple areas indicate lower incident light intensity.): (**a**) System 1; (**b**) System 2; (**c**) System 3; (**d**) System 4; (**e**) System 5. The optimized length of rod *l* is shown in each panel.

Table 3. Lengths and reductions in the absorbed solar power and resonator losses (compared with a 40-mm-long Nd:YAG rod) for the laser rods of the different systems investigated.

System No.	1	2	3	4	5
Laser rod length (mm)	9	18	22	24	27
Absorbed solar power (W)	187.2	202.5	202.5	200.9	199.0
Reduction in resonator losses (%)	77.5	55	45	40	32.5
Reduction in absorbed solar power (%)	22.4	11.9	10.4	9.6	9.4

2.3.4. Step 4: Optimization of Parabolic Surface S2 in the Pump Cavity

The upper edge of parabolic surface S2 was bonded with the lower edge of parabolic surface S1, and the output diameter of the pump cavity was the same as that of the diameter of the quartz tube. The surface equation for S2 was completely determined by the value of the focal length $n/2$. As the focal length must be selected to ensure that the focus of S2 is within the laser rod, all the possible values of n were used to determine the maximum absorbed solar power in the laser rod; the results are summarized in Table 4.

Table 4. Parameters of the parabolic surface S2 and absorbed solar power for the different systems investigated.

System No.	1	2	3	4	5
Focal length of S2 parabola $n/2$ (mm)	6	16	22	32	42
Absorbed solar power (W)	210.6	228.7	232.3	230.7	233.3

The influences of the different parameter optimization steps on the absorbed solar power per unit length of laser rod are summarized in Figure 9. It is apparent that design optimization using TracePro means that these systems are able to efficiently concentrate high-power solar radiation into short lengths of the Nd: YAG laser rod. Thermal effects in the resonator laser and output parameters were further analyzed using the ASLD laser simulation software package, as outlined in Section 3.

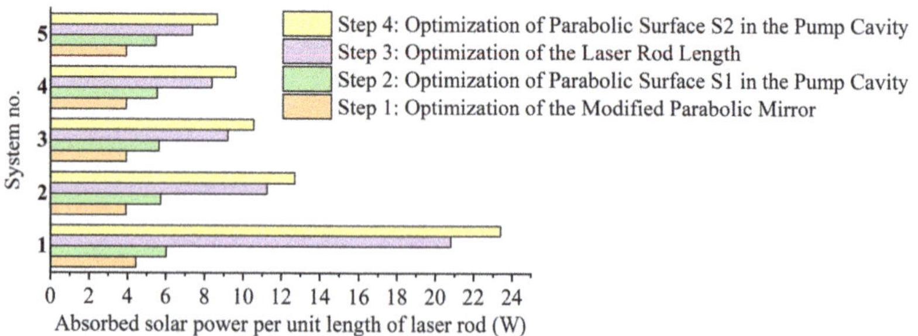

Figure 9. Influence of the optimization processes on the absorbed solar power per unit length laser of rod.

3. Result

To perform laser output power and beam quality analyses, simulated laser resonators were set up in ASLD [28]. In the ASLD simulation, the laser resonator was bounded by a concave mirror with a PR coating and a face with an HR coating ($R > 99.9\%$) at the right and left ends of the laser rod, respectively; the other end face of the laser rod was AR-coated ($R < 0.1\%$) to reduce cavity losses. The length of the laser resonator was 200 mm, and the other losses in the resonator, which occurred mainly at the end faces of the laser rod, were assumed to total to 0.003 cm^{-1}. The "pumping description" in ASLD was imported from the "absorbed–lost flux data matrix" file of the TracePro simulation, which migrated all the solar ray absorption and loss information from the optimized laser medium in TracePro to laser resonators in ASLD. The parameters of the laser medium in ASLD were also consistent with those of TracePro. Moreover, "temperature boundary conditions" of the laser medium were set as cooling with water. In the ASLD analysis, a doping density of 1.39×10^{17} ions/mm^3, fluorescence lifetime of 2.3×10^{-4} s, stimulated emission cross-section of 2.8×10^{-19} cm^2, quantum efficiency of 0.95 [26], and laser medium scattering and absorption coefficient of 3×10^{-3} cm^{-1} for the Nd:YAG medium containing 1.0 at.% Nd^{3+} were used as inputs. A wavelength of 660 nm was used as the mean intensity-weighted absorbed solar pump wavelength [9]. By dividing the laser medium into 169,377 discretization grids and running a finite element analysis (FEA), both thermal conditions and the laser performance could be calculated in ASLD.

Owing to the narrow range of the solar spectrum that can be utilized by the SPL [29], solar power usually contributes to severe heating of the laser medium in SPLs. Therefore, the thermal conditions of the laser rods were analyzed. Figure 10 shows the temperature and thermally induced stress distributions in the laser rods pumped by these concentrator systems, which were evaluated by FEA in ASLD. As the distance between the focus of the parabolic mirror and the central axis b increased, the thermal stress in the laser rod was significantly suppressed, and the maximum temperatures in the rods decreased, from 439.6 K ($b = 0$) to 389.2 K ($b = 20$ mm), 366.7 K ($b = 30$ mm), 354.9 K ($b = 40$ mm), and 353.8 K ($b = 50$ mm). Regarding stress, the maximum values of 118, 86.45, 70.56, and 68.37 N/mm^2, obtained for the systems with b in the range of 20–50 mm, were far below the stress fracture limit for the Nd:YAG medium of 200 N/mm^2 [26], whereas the value for the $b = 0$ system was 189.4 N/mm^2, i.e., close to the fracture limit.

The output beam-quality performance of the laser rods is summarized in Table 5. Using the dual-parabolic-surface pump cavity, solar radiation was coupled into the short laser rods efficiently and uniformly, and high-efficiency conversion from solar power to laser power was realized. The optimized SPL based on the concentrator system with a parabolic mirror having a focal length of 200 mm and offset distance of 30 mm was shown to have the best laser output performance, with an output laser power of 74.6 W and beam quality factors of $M_x^2 = 34.0$ and $M_y^2 = 35.1$. From this result, a laser collection efficiency of 42.2 W/m^2 and a beam brightness figure of merit of 0.063 W were calculated.

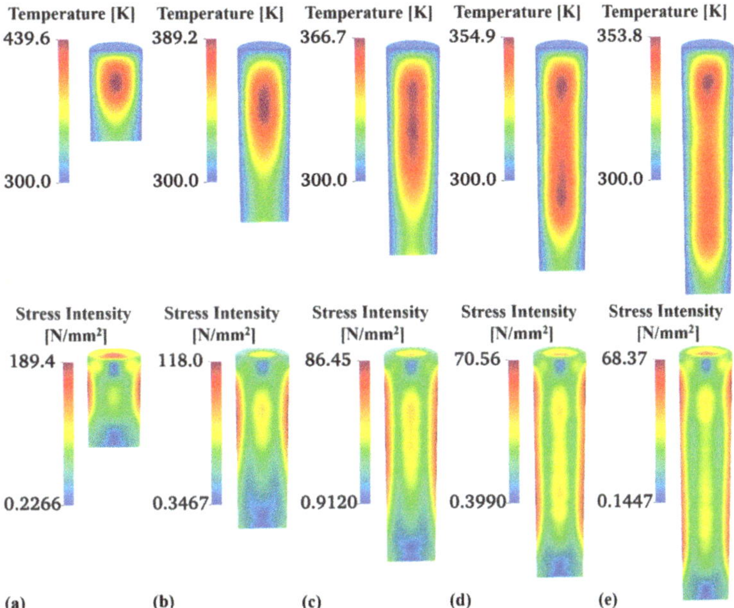

Figure 10. Numerically calculated temperature and stress distributions in Nd:YAG rods in the different optimized solar-pumped laser systems: (**a**) System 1; (**b**) System 2; (**c**) System 3; (**d**) System 4; (**e**) System 5.

Table 5. Output laser performance parameters for different optimized SPL systems.

System No.	1	2	3	4	5
Output mirror reflectivity	0.95	0.94	0.93	0.93	0.93
Laser output power (W)	70.2	73.5	74.6	68.4	71.5
Laser collection efficiency (W/m²)	39.8	41.6	42.2	38.7	40.5
M_x^2	51.8	35.7	34.0	37.1	36.3
M_y^2	53.3	36.9	35.1	39.2	37.1
Conversion efficiency (%)	4.06	4.25	4.31	3.95	4.13

4. Discussion

In this study, a novel compound solar concentrator was proposed and applied to the design of a solar-pumped laser. Significant improvements in not only the solar laser collection efficiency but also the laser beam quality was numerically calculated using this structure combined with a novel dual-parabolic-surface pump cavity. Five systems distinguished by different distances between the focus and central axis of the modified parabolic mirror were modeled in TracePro, and the solar concentrator and pump cavity designs for each of these systems were optimized using TracePro and ASLD numerical analyses. Consequently, it was numerically predicted that by end-side pumping a 5-mm-diameter, 22-mm-long Nd:YAG crystal rod, 74.6 W of continuous-wave laser power with beam quality factors of $M_x^2 = 34.0$ and $M_y^2 = 35.1$ could be obtained, corresponding to a collection efficiency of 42.2 W/m² and a beam brightness figure of merit of 0.063 W, which were 1.1 times and 2.6 times greater, respectively, than the current record values.

A comparison between the results obtained in this study and the previous records, for both end-side-pumping [22] and side-pumping [25] schemes, is presented in Table 6. Although the laser rod was end-side pumped in this work, the beam quality was nearly as high as that previously obtained via side-pumping, and the highly efficient energy conversion characteristic of end-side pumping was retained.

Table 6. Comparisons between results obtained in this work and previous record results for end-side pumping [22] and side pumping [25].

Parameters	Ref. [22]	Ref. [25]	Present Work	Fold Change with Respect to Ref. [22]	Ref. [25]
Collection area (m^2)	1.76	1.71	1.76	-	-
Pumping configuration	end-side-pumping	side-pumping	end-side-pumping	-	-
Laser output (W)	67.8	49.89	74.6	-	-
Collection efficiency (W/m^2)	38.4	29.18	42.2	1.1	1.45
Conversion efficiency (%)	4.0	3.07	4.31	1.08	1.40
Average M^2 factor $\left(\dfrac{M_x^2 + M_y^2}{2}\right)$	55.0	33.0	34.6	-	-
Figure of merit (W)	0.024	0.046	0.063	2.625	1.370

In this study, the most popular solid-state laser medium, Nd:YAG, was used. However, the optical structures developed are applicable for pumping other rod-shaped solar laser media. In particular, it would be interesting to investigate the use of the compound solar collector and dual-parabolic pump cavity with other laser media that absorb more strongly in the solar emission spectrum but suffer from more serious thermal effects, such as Cr:Nd:YAG ceramics, Nd:Ce:YAG, or alexandrite single crystals. The observed characteristics of the developed optical setup, i.e., homogenized pump-light absorption and the suppression of thermal effects, suggest that an SPL design incorporating these more absorbing media with the new optical elements might achieve better performance, i.e., higher laser collection efficiency or better beam quality.

In addition, the structure proposed in this paper was not complex, with only three parts coaxially mounted and, as far as we know, all the specific mechanisms and optical surfaces could be processed. Therefore, the proposed approach could be used for SPLs, which we shall investigate in future.

Author Contributions: Conceptualization, Z.C. and C.Z.; Data curation, Z.C.; formal analysis, Z.C. and Z.Z. (Ziyin Zhao); funding acquisition, C.Z.; investigation, Z.C.; methodology, Z.C. and C.Z.; project administration, C.Z.; resources, C.Z., H.Z. and Z.Z. (Zilong Zhang); software, Z.C. and Z.Z. (Ziyin Zhao); supervision, C.Z., H.Z. and Z.Z. (Zilong Zhang); validation, Z.C., Z.Z. (Ziyin Zhao) and X.Y.; visualization, Z.C.; writing—original draft, Z.C.; writing—review & editing, Z.C., C.Z., H.Z. and Z.Z. (Zilong Zhang). All authors have read and agreed to the published version of the manuscript.

Funding: This research was funded by the National Natural Science Foundation of China, grant number 61378020 and 61775018.

Data Availability Statement: Not applicable.

Conflicts of Interest: The authors declare no conflict of interest. The funders had no role in the design of the study; in the collection, analyses, or interpretation of data; in the writing of the manuscript, or in the decision to publish the results.

References

1. Yabe, T.; Uchida, S.; Ikuta, K.; Yoshida, K.; Baasandash, C.; Mohamed, M.S.; Sakurai, Y.; Ogata, Y.; Tuji, M.; Mori, Y.; et al. Demonstrated fossil-fuel-free energy cycle using magnesium and laser. *Appl. Phys. Lett.* **2006**, *89*, 261107. [CrossRef]
2. Ohkubo, T.; Yabe, T.; Yoshida, K.; Uchida, S.; Funatsu, T.; Bagheri, B.; Oishi, T.; Daito, K.; Ishioka, M.; Nakayama, Y.; et al. Solar-pumped 80 W laser irradiated by a Fresnel lens. *Opt. Lett.* **2009**, *34*, 175–177. [CrossRef]
3. Yabe, T.; Bagheri, B.; Ohkubo, T.; Uchida, S.; Yoshida, K.; Funatsu, T.; Oishi, T.; Daito, K.; Ishioka, M.; Yasunaga, N.; et al. 100 W-class solar pumped laser for sustainable magnesium-hydrogen energy cycle. *J. Appl. Phys.* **2008**, *104*, 083104. [CrossRef]
4. Mori, M.; Kagawa, H.; Saito, Y. Summary of studies on space solar power systems of Japan Aerospace Exploration Agency (JAXA). *Acta Astronaut.* **2006**, *59*, 132–138. [CrossRef]
5. Guan, Z.; Zhao, C.M.; Yang, S.H.; Wang, Y.; Ke, J.Y.; Zhang, H.Y. Demonstration of a free-space optical communication system using a solar-pumped laser as signal transmitter. *Laser Phys. Lett.* **2017**, *14*, 055804. [CrossRef]

6.   Takeda, Y.; Iizuka, H.; Mizuno, S.; Hasegawa, K.; Ichikawa, T.; Ito, H.; Kajino, T.; Ichiki, A.; Motohiro, T. Silicon photovoltaic cells coupled with solar-pumped fiber lasers emitting at 1064 nm. *J. Appl. Phys.* **2014**, *116*, 014501. [CrossRef]

7.   Young, C.G. A Sun-Pumped cw One-Watt Laser. *Appl. Opt.* **1966**, *5*, 993–997. [CrossRef]

8.   Arashi, H.; Oka, Y.; Sasahara, N.; Kaimai, A.; Ishigame, M. A Solar-Pumped cw 18 W Nd:YAG Laser. *Jpn. J. Appl. Phys.* **1984**, *23*, 1051–1053. [CrossRef]

9.   Weksler, M.; Shwartz, J. Solar-pumped solid-state lasers. *IEEE J. Quantum Electron.* **1988**, *24*, 1222–1228. [CrossRef]

10.   Benmair, R.M.J.; Kagan, J.; Kalisky, Y.; Noter, Y.; Oron, M.; Shimony, Y.; Yogev, A. Solar-pumped Er, Tm, Ho:YAG laser. *Opt. Lett.* **1990**, *15*, 36–38. [CrossRef]

11.   Lando, M.; Kagan, J.; Linyekin, B.; Dobrusin, V. A solar-pumped Nd:YAG laser in the high collection efficiency regime. *Opt. Commun.* **2003**, *222*, 371–381. [CrossRef]

12.   Yabe, T.; Ohkubo, T.; Uchida, S.; Yoshida, M.; Nakatsuka, M.; Funatsu, T.; Mabuti, A.; Oyama, A.; Nakagawa, K.; Oishi, T.; et al. High-efficiency and economical solar-energy-pumped laser with Fresnel lens and chromium codoped laser medium. *Appl. Phys. Lett.* **2007**, *90*, 261120. [CrossRef]

13.   Liang, D.; Almeida, J. Highly efficient solar-pumped Nd:YAG laser. *Opt. Express* **2011**, *19*, 26399–26405. [CrossRef] [PubMed]

14.   Dinh, T.H.; Ohkubo, T.; Yabe, T.; Kuboyama, H. 120 watt continuous wave solar-pumped laser with a liquid light-guide lens and an Nd:YAG rod. *Opt. Lett.* **2012**, *37*, 2670–2672. [CrossRef] [PubMed]

15.   Guan, Z.; Zhao, C.; Li, J.; He, D.; Zhang, H. 32.1 W/m^2 continuous wave solar-pumped laser with a bonding Nd:YAG/YAG rod and a Fresnel lens. *Opt. Laser Technol.* **2018**, *107*, 158–161. [CrossRef]

16.   Liang, D.; Vistas, C.R.; Tibúrcio, B.D.; Almeida, J. Solar-pumped Cr:Nd:YAG ceramic laser with 6.7% slope efficiency. *Sol. Energy Mater. Sol. Cells* **2018**, *185*, 75–79. [CrossRef]

17.   Vistas, C.R.; Liang, D.; Almeida, J.; Tibúrcio, B.D.; Garcia, D.; Catela, M.; Costa, H.; Guillot, E. Ce:Nd:YAG side-pumped solar laser. *J. Photonics Energy* **2021**, *11*, 018001. [CrossRef]

18.   Tibúrcio, B.; Liang, D.; Almeida, J.; Garcia, D.; Vistas, C.R.; Morais, P. Highly efficient side-pumped solar laser with enhanced tracking-error compensation capacity. *Opt. Commun.* **2020**, *460*, 125156. [CrossRef]

19.   Liang, D.; Almeida, J.; Vistas, C.R.; Guillot, E. Solar-pumped TEM00 mode Nd:YAG laser by a heliostat—Parabolic mirror system. *Sol. Energy Mater. Sol. Cells* **2015**, *134*, 305–308. [CrossRef]

20.   Mehellou, S.; Liang, D.; Almeida, J.; Bouadjemine, R.; Vistas, C.R.; Guillot, E.; Rehouma, F. Stable solar-pumped TEM00-mode 1064 nm laser emission by a monolithic fused silica twisted light guide. *Sol. Energy* **2017**, *155*, 1059–1071. [CrossRef]

21.   Xiong, S.; He, Y.; Liu, X. Parabolic ring array concentrator for solar-pumped laser. *J. Appl. Opt.* **2014**, *35*, 531–536.

22.   Matos, R.; Liang, D.; Almeida, J.; Tibúrcio, B.D.; Vistas, C.R. High-efficiency solar laser pumping by a modified ring-array concentrator. *Opt. Commun.* **2018**, *420*, 6–13. [CrossRef]

23.   Boutaka, R.; Liang, D.; Bouadjemine, R.; Traiche, M.; Kellou, A. A Compact Solar Laser Side-Pumping Scheme Using Four Off-Axis Parabolic Mirrors. *J. Russ. Laser Res.* **2021**, *42*, 453–461. [CrossRef]

24.   Catela, M.; Liang, D.; Vistas, C.R.; Garcia, D.; Tibúrcio, B.D.; Costa, H.; Almeida, J. Doughnut-Shaped and Top Hat Solar Laser Beams Numerical Analysis. *Energies* **2021**, *14*, 7102. [CrossRef]

25.   Tibúrcio, B.D.; Liang, D.; Almeida, J.; Matos, R.; Vistas, C.R. Improving solar-pumped laser efficiency by a ring-array concentrator. *J. Photonics Energy* **2018**, *8*, 018002. [CrossRef]

26.   Koechner, W. *Solid-State Laser Engineering*, 6th ed.; Springer: New York, NY, USA, 2006.

27.   ASTM G173-03(2012). *Standard Tables for Reference Solar Spectral Irradiances: Direct Normal and Hemispherical on 37° Tilted Surface*; ASTM: West Conshohocken, PA, USA, 2012.

28.   Advanced Software for Laser Design (ASLD). Multiphysics Laser Simulation Software. Available online: http://www.asldweb. com/index.html (accessed on 4 February 2022).

29.   Zhao, B.; Zhao, C.; He, J.; Yang, S. The Study of Active Medium for Solar-Pumped Solid-State Lasers. *Acta Opt. Sin.* **2007**, *27*, 1797–1801.

Article

Fixed Fiber Light Guide System with Concave Outlet Concentrators

Bangdi Zhou, Kaiyan He *, Ziqian Chen * and Shuiku Zhong

School of Physical Science and Technology, Guangxi University, Nanning 530004, China;
1907301114@st.gxu.edu.cn (B.Z.); newzhongshuiku@163.com (S.Z.)
* Correspondence: gredhky@gxu.edu.cn (K.H.); czq8676@163.com (Z.C.)

Abstract: Because a traditional optical fiber light guiding system includes a tracking device, it also inevitably has a complex structure, high construction and maintenance costs, short life span and low reliability. Although several types have been developed for decades, there are no successful products on the market. The biggest cause of the problem is that all traditional optical fiber light guiding systems must have a tracking device. This paper studies a solar fiber optic guide system without a tracking device, hoping to solve this problem. A fixed fiber light guide system using concave outlet concentrators as its receiving unit is proposed. The structure and working principle of the concave outlet concentrator, the receiving unit and the light guide system are introduced. With optical simulation software and the actual sunlight experimental method, this paper first discusses the conceptual design of the concentrator, then studies the transmission efficiency curve of the receiving unit with different angles of incident light, and finally tests the output illuminance of the whole system in actual sunlight. Field test results show that when the average sunshine intensity is about 800 W/m^2, the system has an output of nearly 300 lux at 0.4 m in front of the outlet end of the fiber bundle with only 3.11×10^{-2} m^2 receiving area. This illumination has been able to meet people's daily lighting requirements. The results of computer simulation and actual sunlight experiments show that this fixed optical fiber light guide system with non-tracing structure is feasible. The absence of a tracking structure means that all moving parts of the system are completely discarded. This greatly improves the working reliability and operation life of the light guide system, and greatly reduces the maintenance and operating costs.

Keywords: concentrator; light guide; optical fiber solar system; solar daylighting

Citation: Zhou, B.; He, K.; Chen, Z.; Zhong, S. Fixed Fiber Light Guide System with Concave Outlet Concentrators. *Energies* **2022**, *15*, 982. https://doi.org/10.3390/en15030982

Academic Editor: Jesús Polo

Received: 30 December 2021
Accepted: 25 January 2022
Published: 28 January 2022

Publisher's Note: MDPI stays neutral with regard to jurisdictional claims in published maps and institutional affiliations.

1. Introduction

With the introduction of low-carbon and green lighting concepts in modern society, the lighting mode of deep interior spaces such as basement through solar lighting has attracted extensive attention [1]. At present, there are two main transmission modes on the market to guide sunlight indoors, namely the light tube scheme and the optical fiber light guide scheme. Since the first set of light tube system products was put into application, light tube technology has been relatively mature after decades of development. Recently, researchers have explored more detailed research on the application of light tubes [2–5]. Another scheme, the optical fiber light guide [6], has advantages such as high illumination intensity, flexible light transmission, small space occupancy and so on. As far as this scheme is concerned, there are multiple alternatives; for example, Takashi Nakamura designed an optical waveguide system for solar power applications in space [7,8]. However, there are still no technologically mature products on the market, and the relevant technology is still evolving [9].

At present, the solar concentrators in optical fiber daylighting system are mainly divided into reflective, transmissive and hybrid concentrators. The lighting system using a butterfly solar concentrator [10,11] or trough concentrator [12–14] as its reflective concentrator is a research hotspot. A transmission concentrator mainly includes ordinary convergent lenses and Fresnel lenses. Recently, Lei Li et al. proposed a large Fresnel lens optical fiber daylighting system, and its optical efficiency can reach 11–13% [15]. Vu, N.H. et al proposed an optimized Fresnel lens fiber daylighting system (m-ofds). The system simulation results show that its maximum optical efficiency is 71%. The simulation results also show that the distance that sunlight can be transmitted to the lighting destination is 30 m [16]. For a system with hybrid concentrators, Obianuju et al. used a new two-stage reflective non imaging dish concentrator for collection [17]. Dawei et al. reported on a flexible light guide, which consists of 19 optical fibers and compound parabolic concentrators, and its efficiency can be over 60% [18,19]. In addition, many hybrid optical fiber daylighting and photovoltaic solar lighting systems were also proposed recently [20–22]. However, all these light guiding systems mentioned above require tracking systems with mechanical moving parts and an external power supply. That is a big problem for all traditional optical fiber light guiding systems. The inclusion of tracking devices inevitably results in a complex structure, high construction and maintenance costs, a short life span and low reliability, which is not conducive to market promotion. Efficiency and cost are two important issues in energy application technologies. In general, in solar energy applications, systems with tracking devices are more efficient, while systems without tracking devices are less efficient; the same is true terms of cost. From the history of technology development, the application of tracking devices in large and medium-sized solar systems is commercially reasonable and successful, but the application of tracking devices in small systems, especially small civilian systems, is commercially unsuccessful. Given the endless supply of solar energy, home users are more concerned about maintenance and operating costs, operating reliability and service life. It is easy to understand this issue when you consider the most successful solar product, the solar water heater. Solar water heaters for home use have no tracking device, low cost, long life and are almost maintenance-free. Based on this consideration, in the field of solar light guiding, we also hope to develop a solar light guiding system without a tracking device.

In 2007, a specially designed composite parabolic concentrator was introduced [23]. Subsequently, more in-depth studies were carried out [24,25]. In 2019, based on this concentrator, a fixed optical fiber daylighting system was proposed [26]. For a traditional optical fiber guide system, its shortcomings of high cost, complex structure, low reliability and short life are largely due to the tracking device. However, this fixed system described above completely abandons the tracking device, which would greatly improve the shortcomings of the traditional optical fiber guide system. In this paper, a new structure will be designed to make the fixed fiber light guide system simpler, more reliable and more efficient in higher light energy coupling.

2. The Structure and Working Principle

In the literature published in 2019 [26], a non-tracking light guide system based on the principle of total internal reflection was proposed. The receiving unit of that system that undertakes the solar receiving task is shown in Figure 1a. One of the very important components in this figure is a concentrator with a convex outlet. After further research and analysis on this structure, it was found that this structure still has room for change. Therefore, we propose a new type of receiving unit, as shown in Figure 1b. It can be seen that at least two different designs have been done. Firstly, the optical coupler has been omitted. The benefits are reducing specular reflection and becoming a simpler structure. Secondly, under the same conditions as other geometric parameters of the concentrator, smaller optical fibers can be used to reduce the cost.

The concave outlet concentrator is made of solid block transparent material, as shown in the physical photo in Figure 2a. It realizes light propagation based on the principle of total internal reflection rather than specular reflection. The concave design of the outlet of the concentrator is based on the following reasons. Taking a look at the ray tracing simulation when the bottom end of the outlet is simply made into a plane, as shown in Figure 3a, it can be seen that after the converged exit beam is emitted from the solid concentrator into the air, its convergence angle is large, which is obviously not conducive to the coupling with the optical fiber.

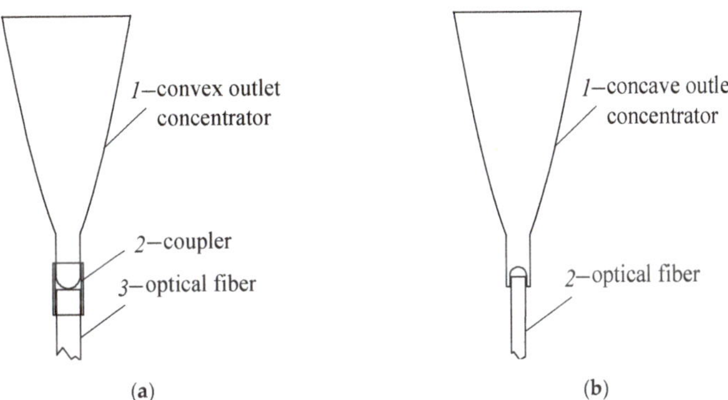

Figure 1. (**a**) Receiving unit with a convex outlet concentrator. (**b**) Receiving unit with a concave outlet concentrator.

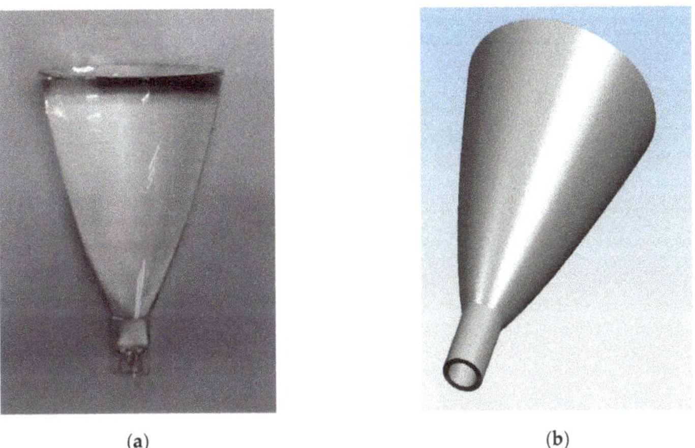

Figure 2. (**a**) Physical photo of the concave outlet concentrator. (**b**) A computer model of the concave outlet concentrator.

Therefore, we changed the flat bottom into a concave spherical structure, which is an arc seen from the outer contour of the shaft section. The outlet end of the concentrator is actually a concave lens. Ray tracing simulation on it is shown in Figure 3b. This shows that the convergence angle of exit beam is smaller than that in Figure 3a, which is good for light transmitting to fiber.

The preliminary design of the concentrator is discussed below. According to the discussion in the literature [24], the geometric structure of the full-size concentrator can be determined as long as the outlet width l and the characteristic parameter k are given (k is defined in the Appendix A). Let $l = 0.014$ m (AB in the Figure 4) according to the available experimental conditions. In order to obtain the optimal k value, we do the following processing: According to the reasonable range of k value discussed in the literature [24], 11 different values of k are determined, and 11 corresponding computer models of the concentrator with slightly different sizes are established. Then, 11 models with the same l value and the same inlet width are respectively imported into the optical software (LightTools) for ray tracing simulation, and their light transmission efficiency was measured. The wavelength of incident light is set to 550 nm, the number of rays to 15,000, and the transmittance of the material to 100%. The results are shown in Figure 4.

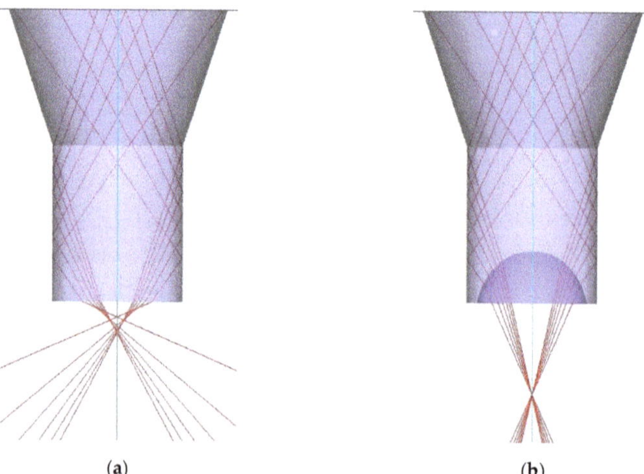

(a)   (b)

Figure 3. (**a**) Ray tracing simulation for a flat outlet concentrator. (**b**) Ray tracing simulation for a concave outlet concentrator.

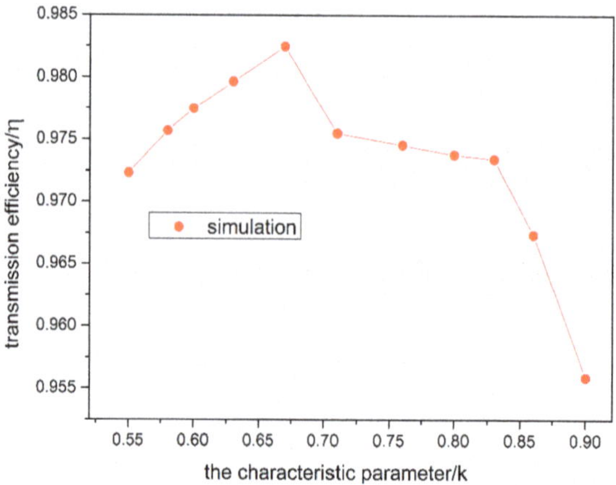

Figure 4. Variation curve of light transmission efficiency with characteristic parameters.

The results correspond to the condition of normal incidence of light and connect with the optical fiber. The light transmission efficiency here is defined according to the following formula [26].

$$\eta = \frac{number\ of\ the\ effective\ emergent\ rays}{number\ of\ the\ effective\ incident\ rays} \tag{1}$$

The effective emergent rays refer to the light rays emitted from the fiber terminal, which is measured by a detection board (a round red board) very close to the end of the optical fiber, as shown in Figure 5. Effective incident rays refer to the light rays entering into the inlet end (receiving surface, or DC in the Figure 6) of the concentrator. It is measured with the help of a detector board at the entrance (a square red board at the entrance of the concentrator in Figure 5). The length and width of this board are larger than the entrance of the concentrator and larger than the cross section of the incident beam. A hole in the middle of the square board has the same area as the entryway of the concentrator. Under the isoplanar condition, the board just covers the entryway of the concentrator. A square column is selected for the incident beam. As long as the light rays do not overflow the square detector board, the effective incident ray number is equal to the ray number of the incident beam minus the ray number on the detection board. Figure 4 shows that there is a maximum value when the k value is between 0.65–0.68. After comprehensive consideration of various factors, the value of k is finally determined to be 0.667. Thus, the values of the two parameters, k and *l*, that determine the concentrator shape are 0.667 and 0.014 m, respectively, which are the geometric parameters of the concentrators used in the experiment.

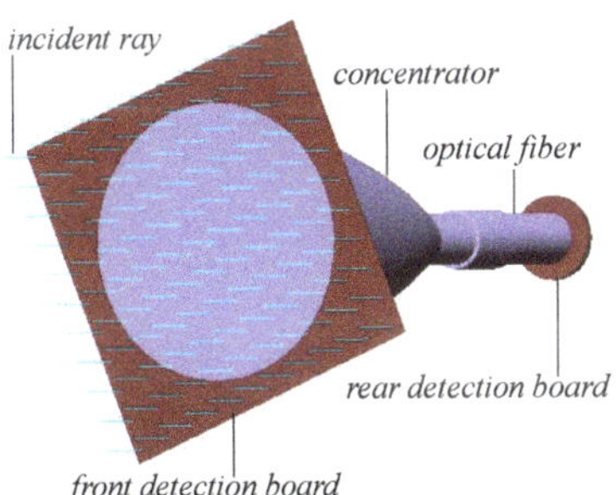

Figure 5. Schematic diagram of ray number detection.

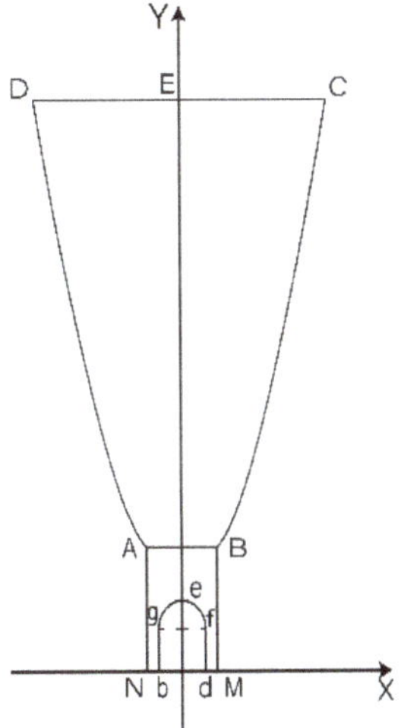

Figure 6. Axial section structure drawing of concentrator.

The detailed axial section geometry of the concentrator is shown in Figure 6. The parabolic CB equation is:

$$y = \frac{1}{18.2}(x + l)^2 \tag{2}$$

where *l* is the length of the line AB. It is the outlet width of the concentrator. The lengths of the lines DC, AB and dM are 66.34 mm, 14 mm and 2 mm, respectively (the experimental model is not a full-scale model. Its height has been truncated by 20%). The segment ef is a quarter arc with a radius of 5 mm. When the curve composed of the six segments EC, CB, BM, Md, df and fe rotates the Y axis, the concentrator would be obtained, as shown in

Figure 2b. The Y axis is its symmetry axis. The calculations of geometric parameters of the front part that is above the focal point of the concentrator are in Appendix A.

3. Results and Discussion

This section is divided by subheadings. It will provide a concise and precise description of the experimental results, their interpretation, as well as the experimental conclusions that can be drawn. It will briefly and accurately describe the experimental process, methods and results, and analyze the results.

3.1. Research on the Transmission Efficiency of the Receiving Unit

To better explore the receiving performance of the concentrator, it was necessary to measure the change of its transmission efficiency under the irradiation of sunlight with different incident deflection angles. The measurement includes computer ray simulation and experiments under actual sunlight conditions. One point to make here is that the concave outlet concentrator is coupled with the transmission optical fiber to form a so-called receiving unit, as shown in Figure 5. The measurement results are shown in Figure 7. Here, we focused on the attenuation characteristic, that is, the shape of the curve, not the specific value of the transmission efficiency. Therefore, the experimental curve was normalized. The transmission efficiency was set to 1 when the incident angle was 0 degrees. The red line in the figure is the results from the computer ray tracing simulation, which uses ray number as an energy unit. The conditions of the simulation are as follows: The transmittance of the concentrator is 100% without considering the absorption loss of the material, and the incident light is monochromatic light with a wavelength of 550 nm. The black line is the result from the actual solar experiment, which uses solar irradiance as the energy unit. In the figure, the angle Φ is the deviation angle from normal incidence.

It can be seen from Figure 7 that the shape of the two curves is similar, and that with the increase in the incidence angle the transmission efficiency gradually decreases. When the incident parallel light rays gradually deviate from the normal, making the propagation process of the light more complex, more light rays cannot be fully reflected into effective output light rays; they go out of the side, as shown in Figure 8. Therefore, the output will decay gradually as the declivity angle of the incident light rays increases. Of course, loss also includes material absorption, which has to do with the wavelength of light, but this loss is hardly reflected in the attenuation curve, because it hardly has much to do with the deflection angle. In other words, different wavelengths will get different efficiencies, but do not affect the shape of the efficiency curve in Figure 7.

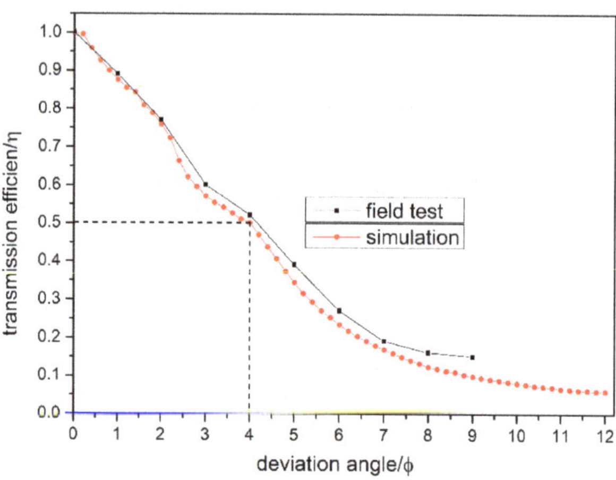

Figure 7. Variation curve of transmission efficiency of a receiving unit with a deflection angle.

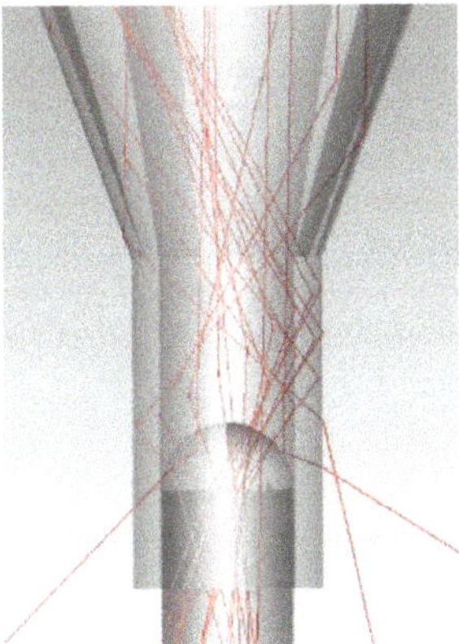

Figure 8. Ray tracing simulation with a small deflection angle.

Another point that needs to be noted is that when the deflection angle become large (more than 8° in figure), the simulation curve still attenuates relatively rapidly, but the experimental curve attenuates relatively slowly. The reason is that sunlight contains both direct light and scattered light. As the deviation angle increases, the loss of direct light gradually increases, and it counts for less and less proportion. However, the scattered light almost stays the same, just taking up more and more proportion. Therefore, the field experimental curve decays slowly at a large angle deflection. However, the incident light corresponding to the computer simulation curve includes only direct light, no scattered light. Therefore, it is still decaying at a relatively fast rate.

However, we should note that the decay is not sudden, but gradual. It can be seen from Figure 7 that when the deflection angle is 4°, the transmission efficiency of the receiving unit is still close to 0.5. That is half of the max transmission efficiency. The value of 4° here is called the full width at half maximum (short for FWHM) of this receiving unit.

Theoretically, if the defects of the device manufacturing and system assembly are not considered, the influence of the deviation of sunlight on the transmission efficiency is axisymmetric in spatial orientation. Therefore, for a fixed receiving unit, the complete light transmission efficiency curve should be as shown in Figure 9. For any receiving unit, the corresponding light transmission efficiency curve has a full width at half maximum, that is, Φ_m in the figure. Specifically, for the model we have established, the FWHM is $\Phi_m = 4°$. This transfer efficiency curve is a very important curve that we rely on to build a fixed optical fiber guide system.

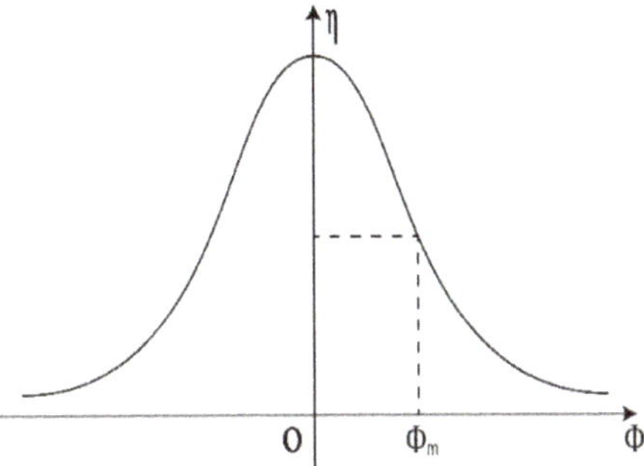

Figure 9. The complete ray transfer efficiency curve of the receiving unit.

3.2. Fixed Fiber Light Guide System

As we can see from the discussion in the previous section, only when the sun moves to the normal incidence position of the receiving unit can the maximum receiving effect be obtained by simply relying on a fixed receiving unit to receive sunlight. Once it deviates from the normal incidence position, receiving efficiency will decrease. The greater the deviation angle, the less the efficiency. However, the feature of the transmission efficiency curve of the receiving unit gives us an inspiration. If we stagger multiple receiving units at certain angles intervals (for example, $2\Phi_m$) on an arc corresponding to the motion of the sun, a fixed light guide system will be gained, as shown in Figure 10.

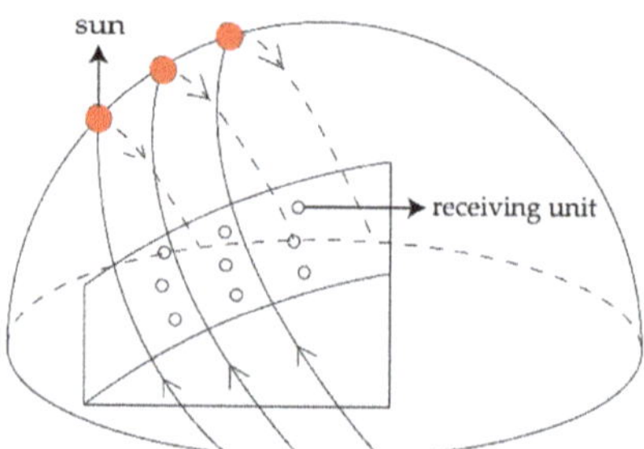

Figure 10. The layout of the receiving unit.

By this arrangement, the resulting fixed light guide system will have a light transmission efficiency curve, which is the staggering superposition of the transmission efficiency curves of nine receiving units, as the red line shows in Figure 11. Within a certain receiving angle range (determined by the number of receiving units), the efficiency curve will be an approximate horizontal line, which is not different from the light guide system with a solar tracking device. It can be seen from Figure 11 that the shape of the curve in Figure 9 is very important. The curve with a large Φ_m is what we want to get. With a larger Φ_m, the design goal can be achieved with fewer receiving units at the same receiving angle. According to the available experimental conditions, we can get a system including nine receiving units, as seen in Figure 12. The angular interval between two receiving units is equal to $2\Phi_m = 8°$. Of the range between $-32°$ to $32°$, the light transmission efficiency decreases rapidly. Based on the same idea, similar problems coming from the seasonal change can be solved by a multi-row arrangement in the direction of the sun's altitude angle, as shown in Figures 10 and 12.

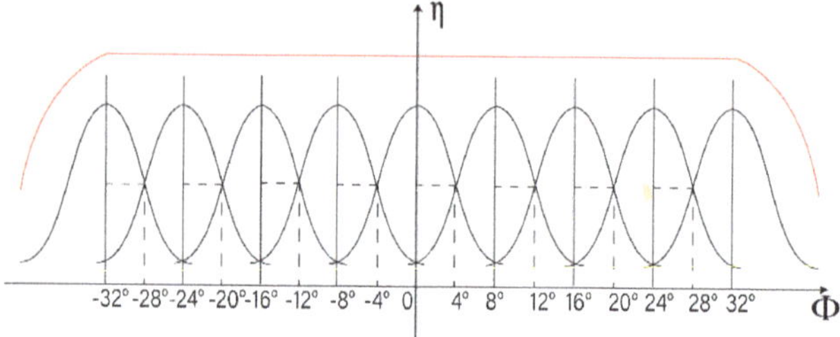

Figure 11. Light transmission efficiency curve of a fixed light guide system.

Figure 12. Fixed light guide system.

3.3. Test on the Fixed Fiber Light Guide System

The test device is shown in Figure 12. The fiber is a plastic optical fiber with a length of 3 m and a diameter of 0.005 m. The ends of nine optical fibers are bundled into a square column shape, as shown in Figure 13, and then are inserted into a small darkroom ($1.2 \times 1.2 \times 1.8$ m^3), that is located inside a room (not in the field of view in Figure 12). The measurement point of illuminance is located 0.4 m in front of the end of the fiber bundle.

(a)

(b)

Figure 13. Photo of the end of the fiber bundle (a) with auxiliary light; (b) without auxiliary light.

The latitude of the experimental site is 23°. When testing, the center line of the light guide system directs to the solar noon position of the sun. Because the sun moves 1° every 4 min, the range of effective output illuminance received by the fixed light guide system is 64°, which is more than 4 h of sunshine, that is, 2 h before noon and 2 h after noon. The measured data are shown in Figures 14 and 15. The red curve is the irradiance of incident sunlight and the purple curve is the output illuminance.

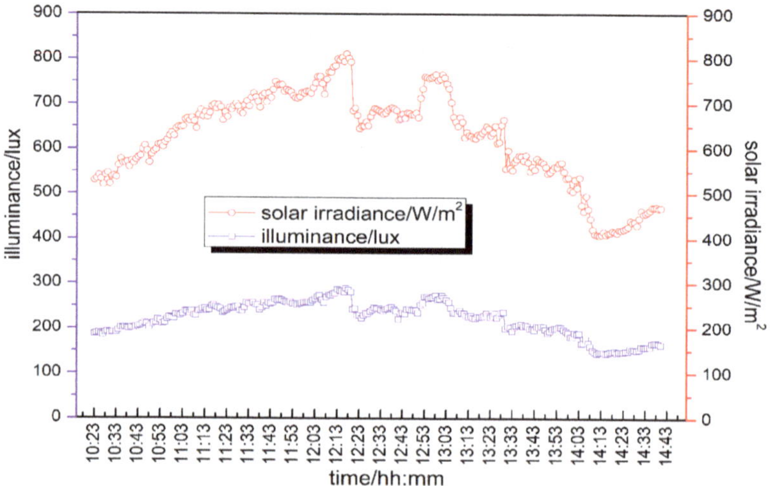

Figure 14. A diagram of the system's output illuminance compared to solar radiation intensity (medium intensity).

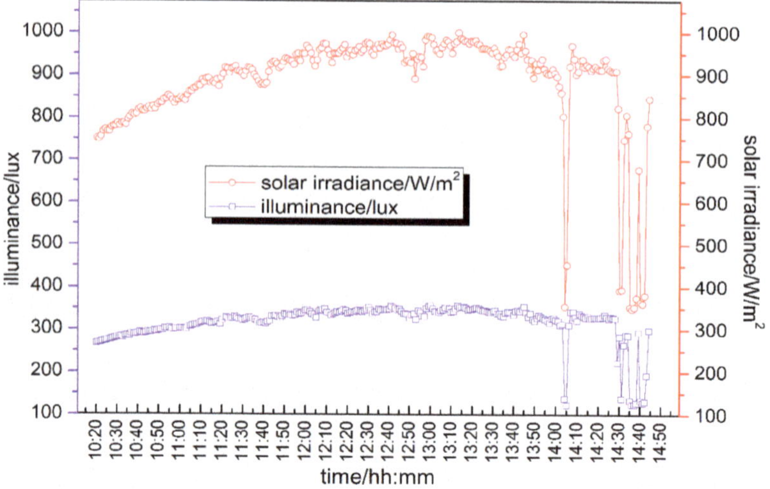

Figure 15. A diagram of the system's output illuminance compared to solar radiation intensity (high intensity).

We can clearly see that the whole output illuminance is almost consistent with the change trend of sunlight irradiance. Without regard to receiving efficiency, the receiving characteristics of this system in receiving sunlight for illumination are the same as the optical guide system with a tracking device. It should be noted that because solar irradiance and illuminance at the end of optical fibers are measured using different instruments that do not have a common time scale system (that is, the two data records are not strictly the same point on the time axis), the change trend of the two curves could not be absolute consistent. The entrance area (solar receiving area) of the nine receiving units is 3.11×10^{-2} m². In the figure, we can also see that when the average intensity of sunlight is about 600 W/m², the output illuminance of the whole system can reach about 200 lux, and when the average intensity of sunlight is about 800 W/m², the output illuminance of the whole system can reach about 300 lux. This illuminance intensity can meet the lighting needs of daily rooms.

From a system architecture perspective, we can improve efficiency and output in the following ways. One way is to shorten the angular interval between the two receiving units. Thus, from Figure 11, we can see that the efficiency curve (red line) will rise. However, it should be noted that if the interval between receiving units remains unchanged, simply increasing the number of receiving units in the same line will not make the efficiency curve rise, but only increase the lighting time. The second way is to increase the number of rows of the receiving unit. The third way is to increase the width *l* of the concentrator outlet (AB in Figure 6), which also increases the width of its entrance (CD in Figure 6). Therefore, more receiving area will be obtained at the entryway of the concentrator. This method can increase the output. Of course, further optimization of the structure and selection of materials with low absorption loss are also good ways to improve the output.

4. Conclusions

A concave exit concentrator and a related fixed light guide system were proposed, and the performance of the concentrator and the whole system were studied and discussed. Due to the ingenious design, the receiving unit composed of this concentrator with a concave exit structure makes the light more favorable for coupling with the fiber. Thus, the receiving unit can dispense with the couplers and can tolerate smaller diameter optical fibers, greatly simplifying the structure of the receiving unit. The fixed light guide system is composed of multiple rows and columns of receiving units, which is similar to a sunflower. Computer ray tracing simulation and field sunlight experiments show the output of the system at different times (within the range of the available receiving time) of each day of the year without a tracking device. The results of the field experimental test show that the system composed of only one row receiving unit with only 3.11×10^{-2} m^2 receiving area had nearly 300 lux output at 0.4 m in front of the outlet end of the fiber bundle under solar irradiance of about 800 W/m^2, and 200 lux at 600 W/m^2. Such illumination can meet the lighting requirements of general rooms, indicating that the design scheme is feasible. This fixed light guiding system that we developed does not have any moving parts. The inevitable benefits are long life and high reliability. Another advantage is that the structure is very simple. Its main composition is the repeated stacking of the simple components shown in Figure 1b. The main material is glass (not counting optical fiber, because optical fiber is also needed for traditional optical guide systems), and the cost is no higher than that of systems with tracking devices. This new system would provide a new feasible scheme for the commercialization of civilian optical fiber guide products.

Author Contributions: Conceptualization, K.H. and B.Z.; methodology, software and validation, K.H., B.Z., Z.C. and S.Z.; formal analysis, investigation and resources, B.Z.; writing—original draft preparation, B.Z. and Z.C.; writing—review and editing, K.H.; visualization, K.H.; supervision, K.H. All authors have read and agreed to the published version of the manuscript.

Funding: National Natural Science Foundation of China, grant number 51868002.

Data Availability Statement: The data that support the study's findings, such as numerical simulation, model or code generated or used during the study, are available upon request from the journal and the corresponding author.

Acknowledgments: This work is supported by the National Natural Science Foundation of China under grant number 51868002.

Conflicts of Interest: The corresponding author declares that there is no conflict of interest on behalf of all authors.

Abbreviations

η	transmission efficiency
k	eigen parameter of the concentrator
Φ	deviation angle
FWHM	full width at half maximum

Appendix A

Appendix A.1. The Formation of Concentrator and the Calculation Formula of Geometric Parameters above the Focal Point

As shown in Figure A1, the two paraboloids with upward openings, F_1 and F_2, are respectively their focal points. Their symmetry axes are parallel to each other, and their equations are, respectively:

$$y = \frac{1}{2p}(x + l)^2, \ (p > 0) \tag{A1}$$

$$y = \frac{1}{2p}(x - l)^2, \ (p > 0) \tag{A2}$$

Coordinates of focus:

$$x_{F_1} = -x_{F_2} = -\frac{l}{2} \tag{A3}$$

$$y_{F_1} = y_{F_2} = \frac{p}{2} \tag{A4}$$

where p is the focal parameter of the parabola and l is the horizontal distance between the focal point F_1 (F_2) and the y-axis, which is also the width of the concentrator outlet AB. Figure A2 is the 3D and 2D structure diagram of concentrator. Line Ab is parallel to the X-axis. Segment AB must satisfy both conditions: (i) It must be positioned above the focus F_1 or F_2 and (ii) its length is exactly the half of the distance.

Between the focal points of the two paraboloids, that is:

$$|AB| = \frac{|F_1 F_2|}{2} = l \ (\text{or} \ |EO| = |OG| = \frac{l}{2}) \tag{A5}$$

Lines AE and BG are perpendicular to the axis, respectively. Take the parabolic segment AD and BC, as well as the straight segment AE and BG, rotating around the axis of symmetry; then a 3D concentrator is obtained, as shown in Figure A2a. If it is translated vertically to the plane *xoy*, a trough shaped combined paraboloid concentrator is obtained, as shown in Figure A2b. The working principle of this concentrator, namely the operating principle of optical path, is shown in Figure A3. Incident light is parallel to the symmetry axis (that is, the y-axis). The light rays reflected from the parabolic segment CB should converge at F_1, but actually converge at point F due to the action of the reflecting mirror AE. The above condition (i) is to ensure that point F is located below the lower opening AB of the parabolic part of the concentrator. Point F is the image of the point F_1 with respect to the plane mirror AE, which happens to be on the symmetry axis of the concentrator. Similarly, the light rays reflected by the parabolic segment AD, then by the reflection of BG, will also converge to the point F. Therefore, point F is the focus of the concentrator.

It can be seen from Figure A3 that ray 2 is already the outermost edge of the ray. The rays to the right of ray 2 will go to DA, instead of propagating down after being reflected by CB. Therefore, the concentrator has an effective minimal width of inlet DC. Tracing ray 2 down, we can see that AE (or BG) also has an effective minimal length.

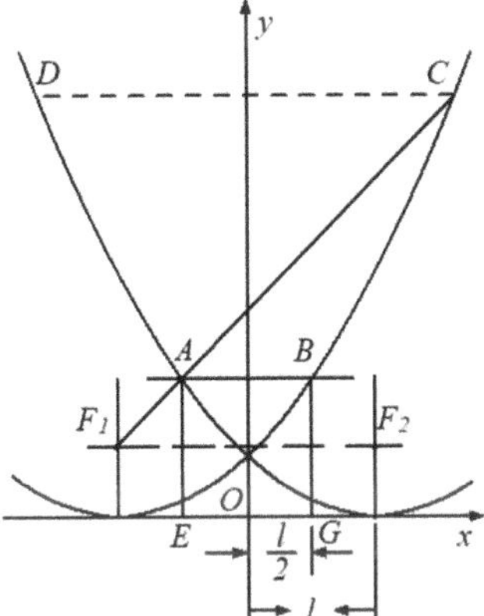

Figure A1. Schematic diagram of the concentrator.

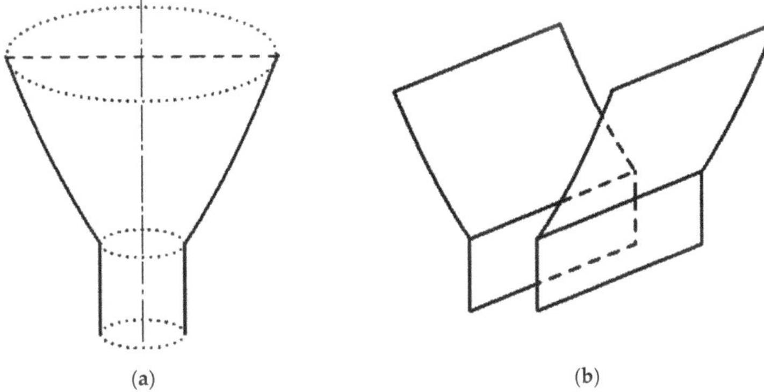

(a)   (b)

Figure A2. (**a**) 3D structure diagram of concentrator; (**b**) 2D structure diagram of concentrator.

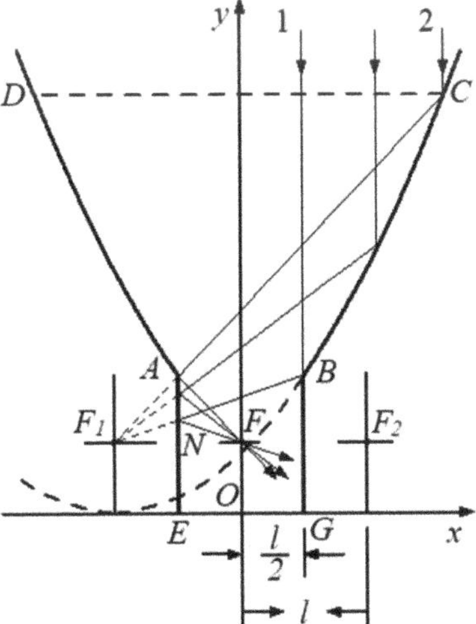

Figure A3. Diagram of light transmission in concentrator.

Appendix A.2. Determination of Main Geometric Parameters of Concentrator

According to the above two conditions that line segment AB must satisfy, and Equations (A1)–(A4), the following equations will be obtained [24]:

The Minimum Length of the Mirror AN

$$h = y_A - y_N = \frac{3l^2}{4p} - \frac{p}{3} \tag{A6}$$

The Effective Diameter of Entrance Aperture

$$\Phi = DC = 2x_C = 2l\left(\frac{5}{4} - \frac{p^2}{l^2} + \sqrt{\left(\frac{9}{4} - \frac{p^2}{l^2}\right)^2 + \frac{p^2}{l^2}}\right) \tag{A7}$$

Diameter Ratio of Inlet to Outlet

$$\beta = \frac{\Phi}{l} = \frac{2x_C}{2x_B} = 2\left(\frac{5}{4} - \frac{p^2}{l^2} + \sqrt{\left(\frac{9}{4} - \frac{p^2}{l^2}\right)^2 + \frac{p^2}{l^2}}\right) \tag{A8}$$

The Ratio of Width to Height for the Tapered Part

$$H = y_C - y_B = \frac{p}{2\frac{p^2}{l^2}}\left(\left(\frac{9}{4} - \frac{p^2}{l^2} + \sqrt{\left(\frac{9}{4} - \frac{p^2}{l^2}\right)^2 + \frac{p^2}{l^2}}\right)^2 - \frac{9}{4}\right) \tag{A9}$$

$$\alpha = \frac{\Phi}{H} = \frac{4\frac{p}{l}\left(\frac{5}{4} - \frac{p^2}{l^2} + \sqrt{\left(\frac{9}{4} - \frac{p^2}{l^2}\right)^2 + \frac{p^2}{l^2}}\right)}{\left(\frac{9}{4} - \frac{p^2}{l^2} + \sqrt{\left(\frac{9}{4} - \frac{p^2}{l^2}\right)^2 + \frac{p^2}{l^2}}\right)^2 - \frac{9}{4}} \tag{A10}$$

The Half Cone Angle of the Outgoing Beam.

$$\tan\theta = \frac{\frac{AB}{2}}{NE - FO} = \frac{\frac{AB}{2}}{(y_A - AN) - FO} = \frac{\frac{p}{l}}{\frac{3}{4} - \frac{p^2}{3l^2}} \tag{A11}$$

The Characteristic Parameter k of the Concentrator.

It can be seen from the above results that the ratio $\frac{p}{l}$ is the only parameter that affects $\alpha\,\beta\,\theta$. It is an important parameter that affects the geometric parameters of concentrator and outgoing beam. It is denoted by k, that is:

$$k = \frac{p}{l} \tag{A12}$$

where k is called a characteristic parameter.

According to the above condition (i), we can get:

$$0 < \frac{p}{l} < \frac{3}{2} \tag{A13}$$

and

$$0 < k < \frac{3}{2} \tag{A14}$$

Then, Equations (A8), (A10) and (A11) become as follows. Their corresponding curves are shown in the Figures A4 and A5:

$$\beta = 2\left(\frac{5}{4} - k^2 + \sqrt{\left(\frac{9}{4} - k^2\right)^2 + k^2}\right) \tag{A15}$$

$$\alpha = \frac{4k\left(\frac{5}{4}k^2 + \sqrt{\left(\frac{9}{4} - k^2\right)^2 + k^2}\right)}{\left(\frac{9}{4} - k^2 + \sqrt{\left(\frac{9}{4} - k^2\right)^2 + k^2}\right)^2 - \frac{9}{4}} \tag{A16}$$

$$\tan\theta = \frac{k}{\frac{3}{4} - \frac{k^2}{3}} \tag{A17}$$

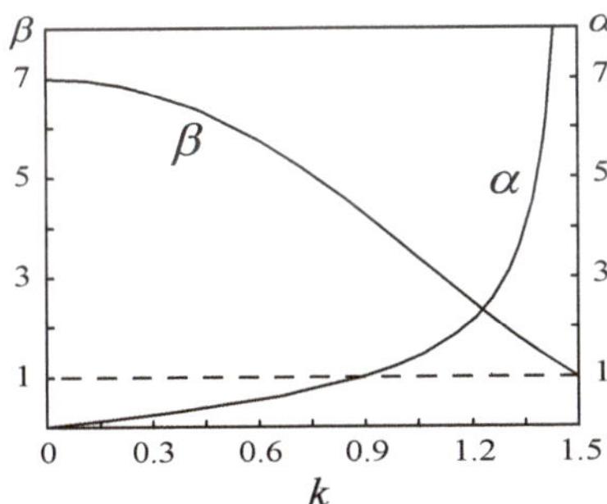

Figure A4. The curves of α and β with respect to k.

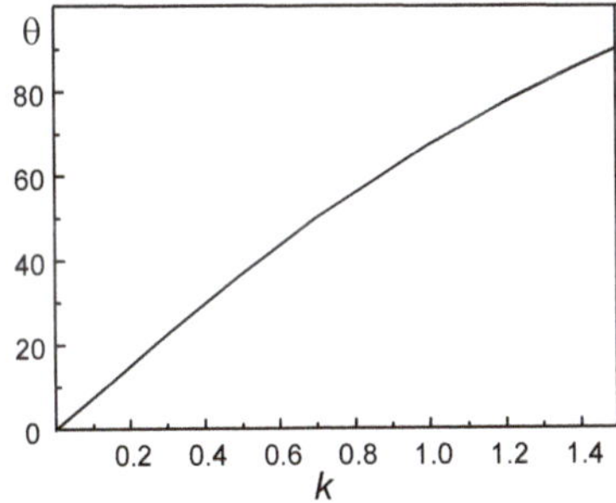

Figure A5. The curves of θ with respect to k.

References

1. Wong, I.; Yang, H.X. Introducing natural lighting into the enclosed lift lobbies of highrise buildings by remote source lighting system. *Appl. Energy* **2012**, *90*, 225–232. [CrossRef]
2. Malet-Damour, B.; Guichard, S.; Bigot, D.; Boyer, H. Study of tubular daylight guide systems in buildings: Experimentation, modelling and validation. *Energy Build.* **2016**, *129*, 308–321. [CrossRef]
3. Petržala, J.; Kocifaj, M.; Kómar, L. Accurate tool for express optical efficiency analysis of cylindrical light-tubes with arbitrary aspect ratios. *Sol. Energy* **2018**, *169*, 264–269. [CrossRef]
4. Darula, S.; Kittler, R.; Kocifaj, M. Luminous effectiveness of tubular light-guides in tropics. *Applied Energy* **2010**, *87*, 3460–3466. [CrossRef]
5. Shuxiao, W.; Jianping, Z.; Lixiong, W. Research on Energy Saving Analysis of Tubular Daylight Devices. *Energy Procedia* **2015**, *78*, 1781–1786. [CrossRef]
6. Dugas, J.; Cariou, J.-M.; Martin, L. Optical fibers and solar energy. *L'Aeronautique Et L'Astronautique* **1982**, *92*, 57–61.

7. Nakamura, T.; Case, J.A.; Senior, C.L.; Jack, D.A.; Cuello, J.L. Optical Waveguide System for Solar Energy Utilization in Space. In Proceedings of the SED2002—2002 International Solar Energy Conference, Reno, NV, USA, 15–20 June 2002.
8. Nakamura, T. Optical Waveguide System for Solar Power Applications in Space. In Proceedings of the SPIE Optical Engineering + Applications, San Diego, CA, USA, 20 August 2009.
9. Feuermann, D.; Gordon, J.M. Solar fiber-optic mini-dishes: A new approach to the efficient collection of sunlight. *Sol. Energy* **2019**, *65*, 159–170. [CrossRef]
10. Kandilli, C.; Ulgen, K.; Hepbasli, A. Exergetic assessment of transmission concentrated solar energy systems via optical fibres for building applications. *Energy Build.* **2008**, *40*, 1505–1512. [CrossRef]
11. Sedki, L.; Maaroufi, M. Design of parabolic solar daylighting systems based on fiber optic wires: A new heat filtering device. *Energy Build.* **2017**, *152*, 434–441. [CrossRef]
12. Ullah, I. Fiber-based daylighting system using trough collector for uniform illumination. *Sol. Energy* **2020**, *196*, 484–493. [CrossRef]
13. Tao, T.; Hongfei, Z.; Kaiyan, H.; Mayere, A. A new trough solar concentrator and its performance analysis. *Sol. Energy* **2011**, *85*, 198–207. [CrossRef]
14. Vu, N.-H.; Shin, S. Cost-effective optical fiber daylighting system using modified compound parabolic concentrators. *Sol. Energy* **2016**, *136*, 145–152. [CrossRef]
15. Lei, L.; Juntao, W.; Zhuodong, Y.; Geng, L.; Kai, T.; Jin, Z.; Jifeng, S. An optical fiber daylighting system with large Fresnel lens. *Energy Procedia* **2018**, *152*, 342–347. [CrossRef]
16. Vu, N.H.; Pham, T.T.; Shin, S. Modified optical fiber daylighting system with sunlight transportation in free space. *Opt. Express* **2016**, *24*, A1528–A1545. [CrossRef]
17. Obianuju, O.N.; Chong, K.-K. High Acceptance Angle Optical Fiber Based Daylighting System Using Two-stage Reflective Non-imaging Dish Concentrator. *Energy Procedia* **2017**, *105*, 498–504. [CrossRef]
18. Dawei, L.; Fraser Monteiro, L.; Ribau Teixeira, M.; Collares-Pereira, M.; Fraser Monteiro, M.L.; Collares-Pereira, M. Fiber-optic solar energy transmission and concentration. *Sol. Energy Mater. Sol. Cells* **1998**, *54*, 323–331.
19. Dawei, L.; Nunes, Y.; Fraser Monteiro, L.; Fraser Monteiro, M.L.; Collares-Pereira, M. 200 W solar energy delivery with optical fiber bundles. In Proceedings of the Optical Science, Engineering and Instrumentation 97, San Diego, CA, USA, 3 October 1997.
20. Yuexia, L.; Longyu, X.; Jinyue, Y.; Jinpeng, B. Design of a hybrid fiber optic daylighting and PV solar lighting system. *Energy Procedia* **2018**, *145*, 586–591.
21. Schlegel, G.O.; Burkholder, F.W.; Klein, S.A.; Beckman, W.A.; Wood, B.D.; Muhs, J.D. Analysis of a full spectrum hybrid lighting system. *Solar Energy* **2004**, *76*, 359–368. [CrossRef]
22. Lv, Y.; Xia, L.; Li, M.; Wang, L.; Su, Y.; Yan, J. Techno-economic evaluation of an optical fiber based hybrid solar lighting system. *Energy Convers. Manag.* **2020**, *225*, 113399. [CrossRef]
23. Kaiyan, H.; Hongfei, Z.; Yixin, L.; Ziqian, C. An imaging compounding parabolic concentrator. In Proceedings of the IESE Solar Word Congress, Beijing, China, 18–21 September 2007; pp. 589–592.
24. Kaiyan, H.; Hongfei, Z.; Zhengliang, L.; Tao, T. A novel multiple curved surfaces compound concentrator. *Sol. Energy* **2011**, *85*, 523–529. [CrossRef]
25. Kaiyan, H.; Hongfei, Z.; Tao, T.; Dai, J. Design and investigation of a novel concentrator used in solar fiber lamp. *Sol. Energy* **2009**, *83*, 2086–2091.
26. Kaiyan, H.; Ziqian, C.; Shuiku, Z.; Yingda, Q.; Haoyue, L.; Junhua, Y.; Bangdi, Z. A solar fiber daylighting system without tracking component. *Sol. Energy* **2019**, *194*, 461–470.

Article

Elliptic Array Luminescent Solar Concentrators for Combined Power Generation and Microalgae Growth

Nima Talebzadeh and Paul G. O'Brien *

Department of Mechanical Engineering, Lassonde School of Engineering, York University, Toronto, ON M3J 1P3, Canada; nimatg@yorku.ca
* Correspondence: paul.obrien@lassonde.yorku.ca

Abstract: The full utilization of broadband solar irradiance is becoming increasingly useful for applications such as long-term space missions, wherein power generation from external sources and regenerative life support systems are essential. Luminescent solar concentrators (LSCs) can be designed to separate sunlight into photosynthetically active radiation (PAR) and non-PAR to simultaneously provide for algae cultivation and electric power generation. However, the efficiency of LSCs suffers from high emission losses. In this work, we show that by shaping the LSC in the form of an elliptic array, rather than the conventional planar configuration, emission losses can be drastically reduced to the point that they are almost eliminated. Numerical results, considering the combined effects of emission, transmission and surface scattering losses show the optical efficiency of the elliptic array LSC is 63%, whereas, in comparison, the optical efficiency for conventional planar LSCs is 47.2%. Further, results from numerical simulations show that elliptic array luminescent solar concentrators can convert non-PAR and green-PAR to electric power with a conversion efficiency of ~17% for AM1.5 and 17.6% for AM0, while transmitting PAR to an underlying photobioreactor to support algae cultivation.

Keywords: luminescent solar concentrator; solar spectrum splitter; power generation in space; microalgae

Citation: Talebzadeh, N.; O'Brien, P.G. Elliptic Array Luminescent Solar Concentrators for Combined Power Generation and Microalgae Growth. *Energies* **2021**, *14*, 5229. https://doi.org/10.3390/en14175229

Academic Editors: Dawei Liang and Changming Zhao

Received: 9 July 2021
Accepted: 19 August 2021
Published: 24 August 2021

Publisher's Note: MDPI stays neutral with regard to jurisdictional claims in published maps and institutional affiliations.

1. Introduction

The progression of human space exploration endeavors depends on extending the duration of crew missions [1–3]. These ambitious missions will require regenerative environmental control and life support systems (ECLSS). Further, it will not always be possible to bring an energy source that can supply the power required for the duration of the mission and onboard energy conversion systems that utilize the energy sources available in space will be needed. Regarding ECLSS for long-term space missions, microalgae cultivation has been investigated as a promising solution owing to its potential to regenerate O_2 from CO_2, high growth-rates and ability to close the carbon loop by providing a source of food [4–6]. Indeed, in the Photobioreactor at the Life Support Rack (PBR@LSR) experiment, an advanced microalgae photobioreactor (PBR) utilizes concentrated CO_2 from a life support rack onboard the International Space Station [7]. Onboard power generation for spacecrafts operating in the inner solar system is often achieved using photovoltaic (PV) solar panels to convert sunlight into electricity. In this work we are interested in the potential use of luminescent solar concentrators (LSCs) that focus sunlight onto PV cells located at their sidewalls for space applications. We are particularly interested in LSCs because, in addition to concentrating sunlight, they can also function as a solar spectrum splitter that separates photosynthetically active radiation (PAR) from non-PAR. Our objective is to design an LSC that splits the solar radiation to simultaneously provide for electric power generation using non-PAR and green-PAR, and microalgae cultivation using PAR, respectively. The ability to provide multiple functions is highly valuable for mass and volume conservation on spacecrafts. Further, using LSCs for PV operation in space can offer additional benefits. For example, small PV cells are located at the edges of the LSC where they can be shielded from harmful radiation [8], and it has been predicted that specific power values greater than 1 kW/kg can be achieved by using LSCs in space [9]. However, the efficiency of LSCs suffers from emission losses, whereby a large fraction of light emitted from the dye within the LSC panel exits the structure rather than being directed toward the edges of the LSC via total internal reflection (TIR). LSCs have a planar configuration, although in this work we investigate the ability of using other structures to achieve TIR over a broad range of angles to minimize emission losses. We show that emission losses can be drastically reduced by structuring the LSC in the form of an elliptic array rather than a planar configuration. In this paper, we design and numerically evaluate the performance of elliptic-array LSCs that can operate in tandem with a PBR and use solar energy to simultaneously generate electric power and provide for microalgae cultivation.

2. Background

Significant research efforts have been undertaken to explore innovative spectral photon management strategies that optimize algae growth conditions and biofuel cultivation systems [10–15]. For example, Sun et al. [16] enhanced microalgae production by embedding hollow light guides in a flat-plate PBR made of PMMA. The hollow light guides also induced turbulent flow, promoting microalgae suspension mixing, and the photosynthetic efficiency of microalgae growth in the PBR was increased by 12.52%. Ooms et al. [17] demonstrated wavelength specific scattering from plasmonic nano-patterned surfaces as a means of addressing the challenge of photon management in PBRs. Modular PBRs were constructed with different reflective substrates including arrays of plasmonic nanodisks, broadband reflectors and untreated glass, and a power efficiency enhancement of 52% was achieved with the plasmonic nanodisk arrays as compared to the case wherein a broadband reflector was used. Furthermore, a 6.5% cyanobacterium growth rate increase was achieved by using a plasmonic substrate in comparison to using a photobioreactor equipped with untreated glass.

As another spectral photon management strategy for optimizing algae-based biofuel cultivation systems, LSCs are one of the most economical and practical options to harvest solar energy because they have the dual-advantage of (1) concentrating light onto a smaller area (resulting in greater geometrical gain) and (2) splitting the solar irradiance into different spectral regions such that it can be used to simultaneously provide power for multiple applications. An LSC is typically comprised of a transparent panel that hosts luminophore molecules. These luminophores, such as quantum dots or organic dyes, absorb incident solar radiation and use this energy to isotropically emit radiation with a longer wavelength. If the photons are emitted within the critical angle, they will be directed, via TIR, toward PV cells located at the edges of the transparent panel. The incident solar radiation outside the spectral region absorbed by the fluorophores may pass through the panel. This technology can be used in a vast range of applications including greenhouse panels, semi-transparent windows that generate electricity or heat [18] and skylights because the transparency, shape, size and color of the panel are easily controlled [19].

The luminophore embedded within the LSCs can be designed to split the solar spectrum into its PAR and non-PAR components and concentrate the non-PAR onto PV cells located at the LSC sidewalls. Over the course of millions of years, photosynthetic organisms have adopted photo-protection mechanisms to naturally regulate energy flow and minimize energy fluctuations within their photocells [20]. Most photosynthetic organisms are highly reflective toward green light, which has the maximum photon flux in the solar radiation spectrum, to avoid overheating. Instead, light harvesting antennas within photosynthetic organisms are often highly absorbing in the spectral vicinity of blue and red light [21]. However, due to several deficiencies and considerable loss mechanisms, the power conversion efficiency of LSCs is still less than 10%. The main operating mechanisms and loss phenomena in LSC devices are described subsequently with reference to Figure 1 [22].

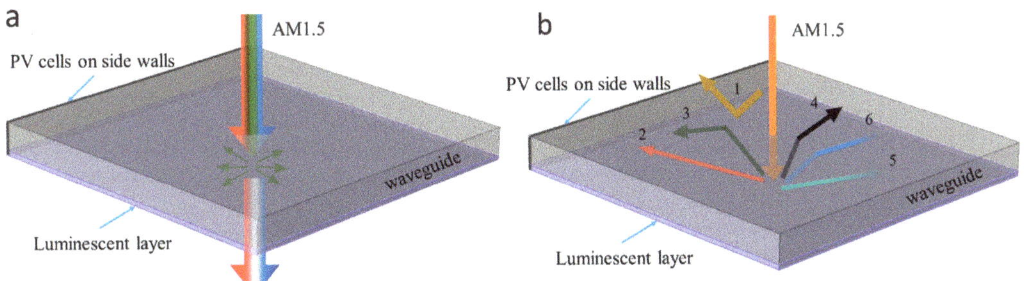

Figure 1. (**a**) Mechanism of an LSC as a solar spectrum splitter and solar concentrator and, (**b**) Operation and loss mechanisms in an LSC: (1) Fresnel reflection loss, (2 and 3) Light guiding of photons emitted outside the escape cone via TIR to PV cells at the sidewalls of the planar waveguide, (4) Emission losses of photons emitted within the escape cone, (5) Internal absorption losses, (6) Surface scattering losses at the waveguide/air boundary.

A main reason for low efficiencies in LSCs is transmission losses because a large amount of incident sunlight passes through the panel. Transmission losses occur because luminophore dyes absorb over a small spectral region compared to the spectrally broad solar irradiance. However, this feature makes it possible to use LSCs as semitransparent windows for numerous applications including greenhouses and building facades. Other loss mechanisms include surface reflection loss (which is ~4% for light incident from the normal direction), emission losses due to light being emitted at angles less than the critical angle for which TIR occurs, and emission losses due to surface roughness, which increases photon scattering out of the LSC, and internal absorption losses, which includes the absorption of light by the host matrix waveguide material and the re-absorption of emitted photons by other dye molecules. Many efforts have been applied to reduce the aforementioned losses in LSCs. The methods of reducing surface losses include fabricating selective mirrors, Bragg reflectors, or plasmonic structures at the LSC surfaces, and aligning the luminophores within the LSC to control the path of the emitted light.

Efforts to reduce internal losses caused by re-absorption include studies on novel organic fluorescent dyes, quantum dots and inorganic phosphors [23]. Correia et al. [24] provided a comprehensive overview about the potentials of lanthanide-based organic-inorganic luminescent dyes in order to increase the performance of LSCs. Meinardi et al. [25] designed and fabricated CdSe/CdS quantum dots that utilize a large Stokes shift to eliminate re-absorption losses in large-area LSCs and optical efficiencies exceeding 10% with a concentration factor of 4.4 were achieved. Buffa et al. [26] developed dye-doped polysiloxane rubber waveguides for LSC systems and evaluated the potential of enhancing fluorophore fluorescence efficiency in the presence of different concentrations of Au nanoparticles. Unlike the more common waveguide hosts (e.g., PMMA), polysiloxane rubber is flexible, which enables applications in tents, fabrics, sleeping bags or other such devices that can be rolled up, folded or otherwise deformed. The first transparent near-infrared (NIR)-absorbing LSC with high transparency was demonstrated by Zhao et al. [27]. In this study, they developed luminophore blends of cyanine and cyanine salts and synthesized cyanine salt-host blends with quantum efficiencies greater than 20% accompanied with spectrally selective NIR harvesting. Debije et al. [28] increased the energy output of LSC waveguides by adding white scattering layers to the bottom side of LSCs separated from the waveguide by an air gap. Cambié et al. [29] developed a Monte Carlo ray tracing algorithm to simulate photon paths within LSC-based photomicroreactors and experimentally validated their results. Chou et al. [30] fabricated a flexible waveguiding LSC that exhibits high optical efficiencies and great mechanical flexibility. In this research, a certified power conversion efficiency (PCE) of 5.57%, with a projected PCE as high as approximately 18% was reported. A significant amount of research has also been performed to investigate the performance of LSCs for agrivoltaics, algae cultivation, basil growth and greenhouse applications [11,31–36]. For example, Detweiler et al. [11] fabricated a wavelength selective LSC panel that harnesses the green light portion of the solar irradiance, most of which is not used for algae growth. The LSC panel contains Lumogen Red 305 dye that absorbs and emits green (~580 nm peak) and red light (~620 nm peak), respectively. The fluorescently emitted red light was either used to enhance algal growth, or waveguided and captured by PV cells to be converted into electricity. The results revealed that the microalgae growth rates under the LSC panels were equivalent to the growth rates under the full solar radiation spectrum.

Based on Snell's law, all the photons emitted from luminescent dyes within an LSC that impinge on the surface of the LSC panel at an angle smaller than the critical angle will be emitted to the surrounding medium (loss four in Figure 1b), which is one of the main loss mechanisms in LSCs. The measurements reveal that the accumulated emission losses, considering secondary absorption and emission events and scattering due to surface roughness can reach 50–70% [37]. In this context, all the strategies for reducing surface losses are based on the following two main processes: integrating selective mirrors and aligning the luminophores to directionally control the emission of light [38,39]. However, topology-based strategies wherein the shape of the host medium is altered and optimized to reduce emission losses have yet to be explored. This work evaluates the benefits of shaping the medium hosting the dye in an elliptic-based, rather than planar-based, configuration to minimize the absorption and surface (including scattering and cone zone) losses for LSCs. The performance of these novel elliptic-based LSCs is investigated and compared to conventional LSCs for the dual application of simultaneously providing power for algae and PV electricity production systems.

3. Description of the Elliptic Array Solar Spectrum Splitter and Simulation Methods

In this study, we present a novel spectral-splitting solar concentrator (SSS) that has an elliptical configuration that partitions the solar irradiance into photosynthetically inactive radiation (non-PAR) and green PAR, which is typically not strongly absorbed by microalgae, to power PV cells and blue and red PAR to be utilized in microalgae cultivators.

An ellipse has two focal points, and any ray emitted from one of its focal points that is specularly reflected from its surface will be directed toward its second focal point (Figure S1 in the Supplementary Material, Section A). In the proposed SSS, we make use of this ability of an ellipse to transfer light rays emitted from one of its focal points to its second focal point to design an LSC panel with minimal optical losses. The shared focal point in the proposed SSS is occupied with the concentrated luminescent dye that absorbs green PAR or non-PAR and re-emits radiation at an equal intensity in all directions. The light radiated from the luminescent dye is directed toward the focal points of the adjacent unit cells.

This phenomenon, whereby the light emitted from one focal point is refocused at the focal points of the adjacent unit cells, is achieved by designing the curvature of the ellipses such that the radiation emitted at their shared focal point undergoes TIR at their curved sidewalls. If the medium internal to the SSS has an index of refraction equal to n_1 and the medium external to the structure is air ($n_{air} = 1$) then, according to Snell's law, the radiation emitted from the focal point of the overlapped ellipses will undergo TIR at the surface when the following Equation (1) is satisfied:

$$\frac{c}{a} > \tan\left(\sin^{-1}\left(\frac{1}{n_1}\right)\right) \tag{1}$$

The minimum value of c/a for which the radiation emitted from the focal point of the ellipse will undergo TIR at the ellipse surface is plotted as a function of n_1 (the medium internal to the ellipse) in Figure 2b for the case in which the external medium is air. For example, if the ellipse was made from glass ($n_{glass} = 1.49$), polymer ($n_{high\ refractive\ index\ polymers} = 1.7$) or titanium dioxide ($n_{TiO2} = 2.5$), the minimum values of c/a that would satisfy the TIR condition defined by Equation (1) would be 0.905, 0.727 or 0.436, respectively. Based on the definition of an ellipse $c/a < 1$, which implies that the refractive index of the medium internal to the ellipse must be 1.41 or greater, as indicated by the red line in Figure 2, to satisfy the TIR condition if the medium external to the ellipse is air. The inset at the top right in Figure 2b shows five different shapes of ellipses over the range of $0.905 < c/a < 1$. The TIR condition defined by Equation (1) for the case in which the internal medium is glass is illustrated in Figure 2c, which shows that the radiation emitted from the shared focal point of two overlapping ellipses escapes the SSS structure for values of $c/a < 0.905$; conversely, all the rays emitted from the focal point undergo TIR at the surfaces of the glass-based ellipse for the case in which $c/a \geq 0.905$.

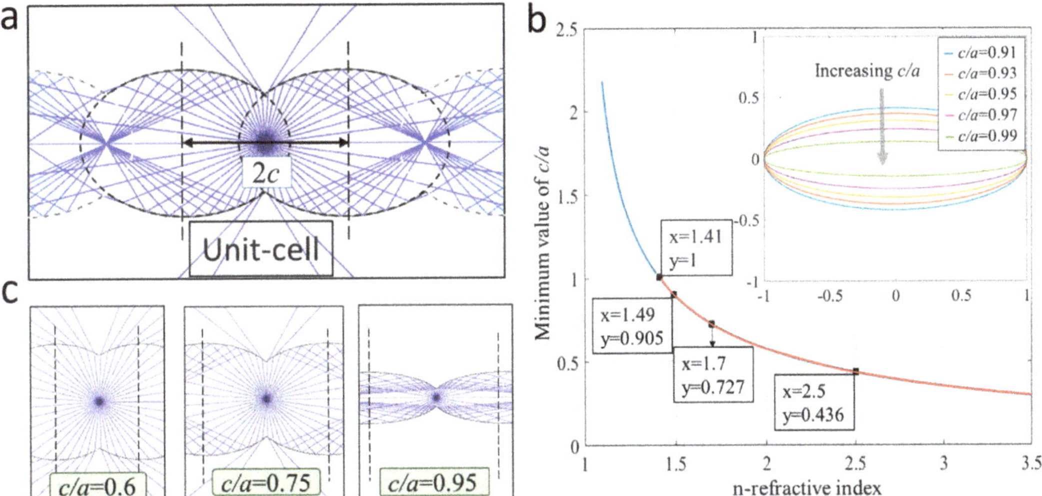

Figure 2. (a) A unit cell that hosts concentrated luminescent dye at the shared focal point of two ellipses. The width of the unit cell is 2C. (b) The minimum value of c/a for which TIR will occur for radiation emitted from the focal point plotted as a function of the refractive index of the medium internal to the ellipse (assuming the medium external to the ellipse is air) and (c) propagation behavior of radiation emitted from a point source located at the shared focal point of overlapping ellipses for the cases in which the index of refraction of the internal medium is $n = 1.49$ and the ellipse parameters are $c/a = 0.6$, $c/a = 0.75$ and $c/a = 0.95$.

Herein, we use the ellipse-based unit cell shown in Figure 2 in the design of a spectral-splitting solar concentrating panel that transmits blue and red PAR, such that it can be used for algae cultivation while absorbing green PAR and non-PAR incident from the solar irradiance to generate electric power. A cross-section of this panel, in the "X–Y" plane, is shown in Figure 3a, and the shape of this panel is realized in three-dimensions by translating this cross-section in the "Z" direction, as shown in Figure 3b. For conciseness, we refer to this panel as an elliptic array solar spectrum splitter (EASSS). Additionally, shown in Figure 3b is an array of Petzval lenses that focuses the incident solar irradiance onto the focal lines within the EASSS. The concentrated light beams coming from the Petzval lens array are incident onto an array of linear receivers (negative lenses) residing on the upper surface of the EASSS, which facilitate the coupling of the incident light onto the focal lines. The incident green PAR and non-PAR is absorbed by luminophores concentrated along the focal lines within the EASSS, while the blue and red PAR is transmitted and leaves the EASSS through a linear exit port located beneath the focal line. The radiation from the luminophores, emitted with equal intensity in all directions, is confined within the EASSS via TIR and is directed toward the PV cells embedded at the side walls of the EASSS. In this case, luminescent dye absorbs the spectral region from ~500–600 nm and from ~740–1100 nm and re-emits photons at wavelengths matched to the single junction PVs at the EASSS side walls. For example, Lumogen Red 305, one of the most commonly used dyes in LSCs, which has an absorption peak of ~580 nm and an emission peak of ~620 nm, and CY (a cyanine derivative), with absorption spectra peaks at 742 nm and NIR emission peaks at 772 nm, can be utilized. InGaP ($E_{bandgap}$ = 1.82 eV), single junction Si ($E_{bandgap}$ = 1.1 eV) and GaSb ($E_{bandgap}$ = 0.67 eV) solar cells can be used as PV cells at the panel sidewalls to harvest different ranges of the non-PAR spectrum. The part of the incoming beam that is not absorbed by the luminescent dye can be used to cultivate algae in a photobioreactor located beneath the EASSS [40–47].

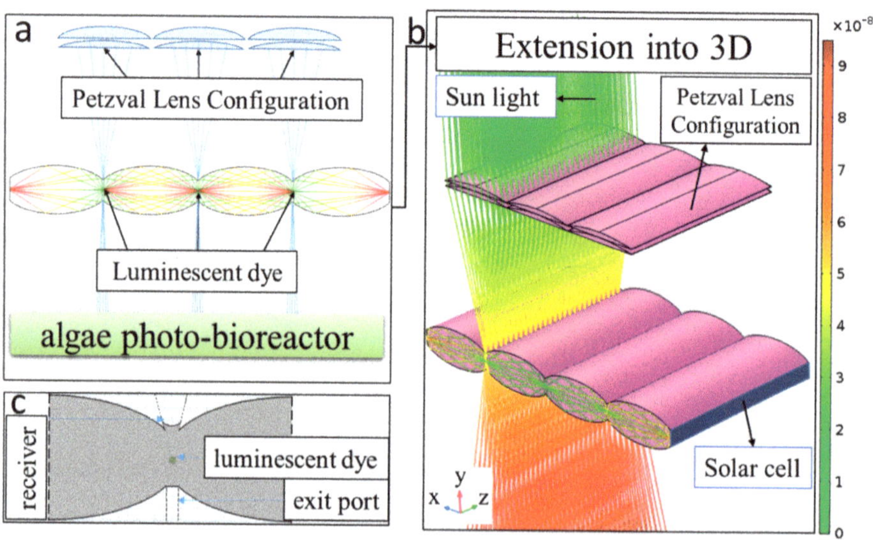

Figure 3. (**a**) The proposed configuration of Elliptic Array Solar Spectrum Splitter (EASSS) in 2D; (**b**) extension of the EASSS into 3D and; (**c**) The concept of receiver to couple the incoming concentrated light from Petzvel lens to the luminescent dye area and send out through the exit port.

In conventional flat-panel LSCs, the light emitted within a broad cone does not undergo TIR at the panel surface and emission losses are ~25%. In the EASSS, TIR occurs for all the light emitted from the focal line with the exception of the light that is emitted onto the receiver or the exit port. As discussed subsequently with reference to Figure 3, the expected emission losses from the EASSS are just a few percent. The surface scattering losses can be estimated using the equation $(1 - D)^N$, where D is the surface scattering coefficient and N is the number of ray collisions with the surface of the EASSS. Moreover, the internal absorption losses can be determined using the expression $e^{-\alpha L}$, where α is the absorption coefficient of the internal medium the EASSS is comprised of, and L is the trajectory path length of the rays through this internal medium.

In the literature, Monte Carlo ray-tracing methods have been used to model LSCs [46–48]. Herein, we numerically model the path length, emission losses and optical efficiency for the light emitted from the focal line inside the EASSS using COMSOL Multiphysics software (version 5.4) ray optics and heat transfer modules supplemented with MATLAB for analytical analysis. The number of rays emitted from each point source was a minimum of 10^6, and the mesh sizes are normalized to the focal distance. The non-dimensional mesh size does not exceed 0.01, and the maximum relative tolerance for convergence is 10^{-5}.

The analysis is performed as follows: Firstly, an EASSS structure without entrance ports (receivers) or exit ports is considered. Specifically, surface scattering losses, in terms of the number of surface scattering events, N, within the EASSS is modeled and analyzed in 2D and compared with that of the conventional flat-panel LSC. Then, the absorption losses, in terms of the mean light-ray path length, are determined for the EASSS and flat-panel LSC. Subsequently, the effect of structure thickness on the absorption and surface scattering losses are investigated. Afterward, the receivers and exit ports are introduced on the upper and lower sides of the EASSS, respectively. The effect of altering the structure of the receiver on the performance of the EASSS is considered and discussed thereafter. In practice, all the luminescent dye required to absorb the incident non-PAR spectra would occupy a finite volume and could not lie precisely on the focal lines within the EASSS. In this regard, the effects of the finite volume occupied by the dye and the sensitivity of the performance of the EASSS with respect to the displacement of the dye from the focal line are investigated. Furthermore, considering the highly concentrated solar irradiation on the narrow receivers, thermal analysis of the EASSS is carried out. Finally, the optical and power conversion efficiency for a sample case of an EASSS working in conjunction with a photo-bioreactor is analyzed and discussed.

4. Results

4.1. Surface Scattering Losses in the EASSS

The number of scattering events per meter width in the x-direction is plotted as a function of the height ($2b = h$) and c/a ratio of the EASSS in Figure 4. For a constant height of the EASSS structure, as c/a increases the number of unit cells (focal lines) per meter within the EASSS panel decreases and, therefore, the number of surface collisions decreases, which lowers the surface scattering losses. In general, as c/a and h increase, surface scattering losses decrease.

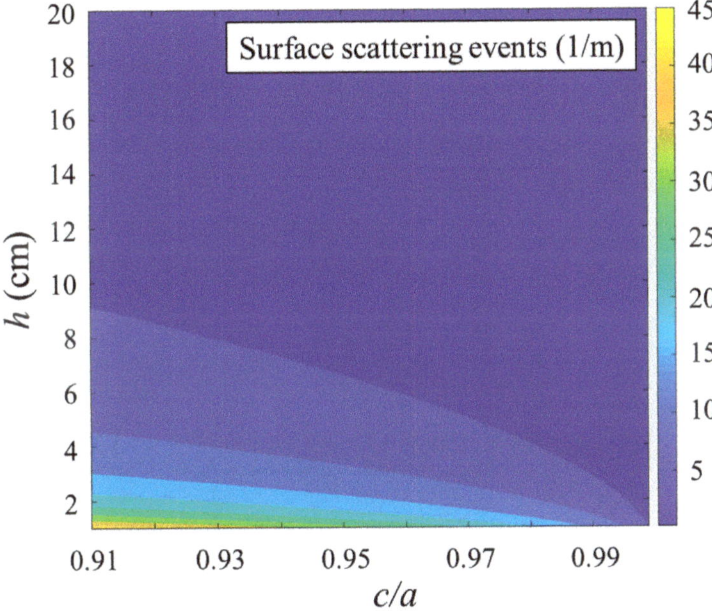

Figure 4. Number of surface scattering events per meter, within an EASSS panel plotted as a function of h and c/a.

The number of surface scattering events per meter in the x-direction is plotted as a function of LSC height for an EASSS with $c/a = 0.91$ and a conventional flat-panel LSC in Figure 5a. Figure 5b shows comparisons of the trajectory of the light emitted from a point source located at the center of the EASSS to the trajectory of the light emitted from the center of a conventional LSC panel. When the panel height is 2 cm, the number of surface scattering events in the EASSS and conventional LSC panels are $N = 22/$m and $N = 55/$m, respectively. When the panel height is increased to 10 cm, the number of surface scattering events in the EASSS and conventional LSC panels is $N = 4\ \text{m}^{-1}$ and $N = 11\ \text{m}^{-1}$, respectively. Thus, the number of surface scattering events in the EASSS is typically less than half that in the convectional LSC, which helps to minimize surface scattering losses.

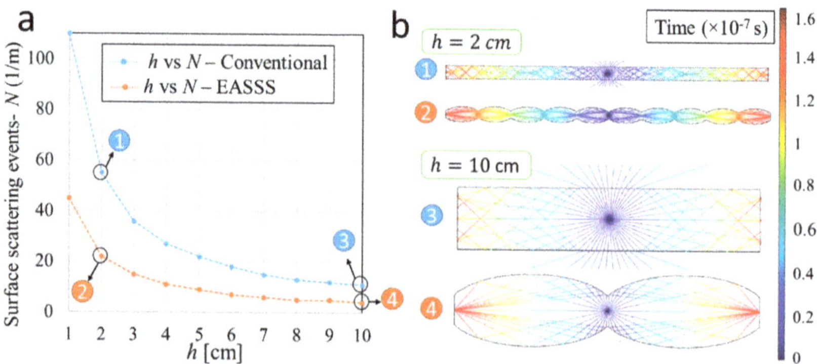

Figure 5. (**a**) A comparison of surface scattering loss events on a per-meter basis for light propagating along the planar direction (the *x*-direction) of an EASSS and a conventional flat-panel LSC for a constant value of $c/a = 0.91$; and (**b**) The trajectory of light emitted from the center of a conventional LSC panel and EASSS for two cases wherein the height of the panels is $h = 2$ and $h = 10$ cm.

4.2. Absorption Losses in the EASSS

It can be noted that absorption losses are independent of structure height, h, because, for a constant value of c/a, as h increases the optical path length remains invariable. On the other hand, as c/a increases, the curvature of the EASSS surfaces increase and the light emitted from the focal lines is more strongly directed toward the EASSS sidewalls, resulting in a decrease in the optical path length.

The optical path length of radiation emitted from the center of a panel to its sidewalls is shown in Figure 6a for the EASSS and the conventional LSC panels for a height of $h = 2$ cm and a width of 2 m. For a point source located in the middle of a conventional LSC, the emission at the critical angle (*C.A*) (~42° when the index of refraction of the LSC is 1.49) and *C.A/2* results in optical path lengths of 150 and 110 cm, respectively. For the EASSS, however, the optical path length is independent of the emission angle. Furthermore, the optical path length ranges from 110 cm (for $c/a = 0.905$) to 100 cm (for $c/a = 0.999$). As can be seen in Figure 6b, the number of unit cells (focal lines) in the EASSS increases as c/a decreases, resulting in a longer optical path length for an EASSS with a fixed height and width of 2 cm and 2 m, respectively.

The optical path length and number of surface scattering events are plotted as a function of the height of the EASSS panel, h, in Figure 7. As h decreases, the number of surface scattering events increases, thereby increasing the expected surface losses ($c/a = 0.91$).

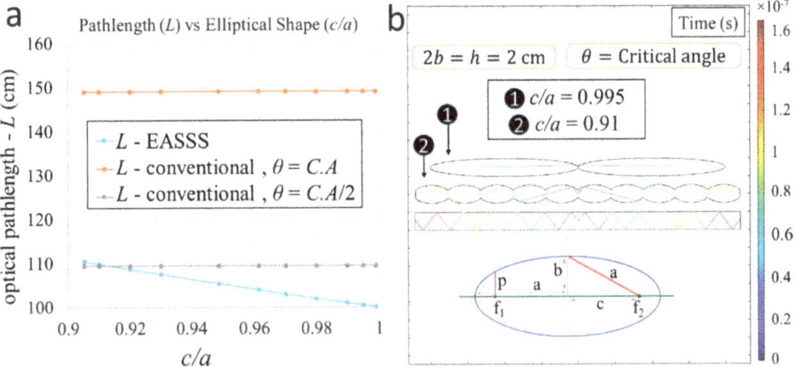

Figure 6. (**a**) Optical path lengths of radiation emitted from the center of the EASSS and conventional LSC panels for a structure with a width of 2 m. For the conventional LSC, the path lengths for emission at the critical angle for which TIR occurs (θ = C.A.) and for emission at half the critical angle are considered; and (**b**) Geometrical illustration of the ray propagation in conventional LSC and EASSS with c/a = 0.91 and 0.995 for h = 2 cm.

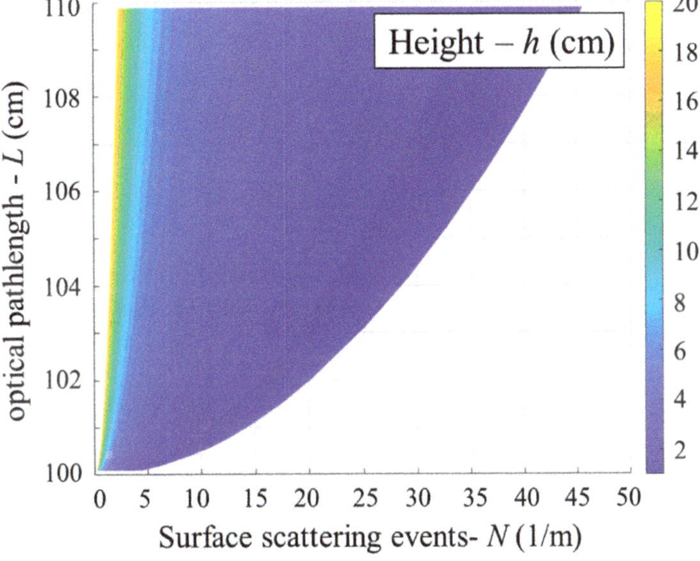

Figure 7. The relationship between the EASSS thickness (h) and the optical path length and the number of surface scattering events for light emitted from the focal lines within the EASSS.

4.3. Transmission Losses in the EASSS

Figure 8 shows the fraction of radiation emitted from luminophores at the centers of the EASSS and conventional LSC panels that are incident onto PV cells located at the panel sidewalls as a function of time. In this figure, two EASSSs with c/a = 0.995 and c/a = 0.91, and a conventional LSC, are considered, each having a height of h = 2 cm. The distance between the luminophore at the center of the panels and the PV cells at the panel sidewalls is assumed to be one meter. As can be seen in Figure 8, for the EASSS, the emitted photons from luminescent dyes reach the solar cells sooner, because they have a shorter distance to travel, in comparison to the conventional LSC. The shorter distance and duration for photons propagating in the EASSS, as compared to the conventional LSC, translates to a reduction in transmission ($\eta_{Transmission}$), self-absorption ($\eta_{Self\text{-}abs}$), surface roughness ($\eta_{Roughness}$) and emission ($\eta_{Emission}$) losses, which will be discussed subsequently.

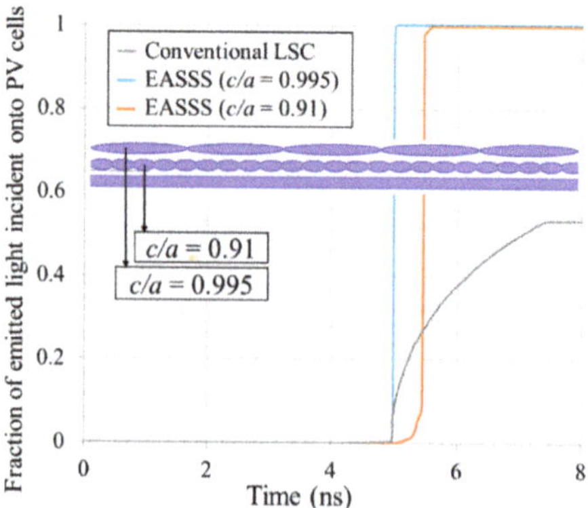

Figure 8. Radiation emitted from the center of the EASSS and conventional LSC panels incident onto the PV cells at the panel sidewalls plotted as a function of ray travelling time for $h = 2$ cm.

4.4. Optical Receiver Design

As mentioned previously, a linear array of Petzval lenses is used to collimate the solar irradiation into the EASSS along the focal lines where the luminescent dyes are concentrated. The Petzval lens configuration is composed of two positive lens groups separated by an air gap. To enable the incoming concentrated blue and red PAR to efficiently exit the underside of the EASSS, a negative lens element is placed on top of the entrance port on the upper side of the EASSS, as shown in Figure 3c. The entrance port introduces emission losses, which are discussed subsequently.

4.4.1. Emission Losses as a Function of the Optical Receiver Design

It is clear that increasing the width of the optical receiver port (O), as shown in Figure 9c, results in higher emission losses. For constant values of O and h, as c/a increases, $p = b^2/a$ decreases and emission losses increase. From the perspective of structure thickness (h), for constant values of O and c/a, as h increases, p increases and emission losses decrease. The emission losses are shown as a function of O and c/a for three different values of $h = 2$, 5 and 10 cm in Figure 9a–c, respectively.

4.4.2. Emission Losses as a Function of the Volume Occupied by the Luminescent Dye in the Vicinity of the EASSS Focal Line

The results presented in all the previous sections were calculated in 2D under the assumption that the luminescent dye is located precisely along the focal line of the overlapped ellipses. In practice, the dye molecules will occupy a limited volume, the size of which depends on the amount of dye required to absorb the incident green PAR and non-PAR solar photon flux. Figure 10a,b show the percentage of radiation emitted from the focal line that is lost through the upper and bottom entrance and exit ports as a function of the in-plane (X–Z plane shown in Figure 2b) and out-of-plane (Y–Z plane shown in Figure 3b) displacement of the luminescent dye from the focal line for two sample EASSS configurations with $c = 5$ cm and $h = 2$ and $h = 4$ cm, respectively.

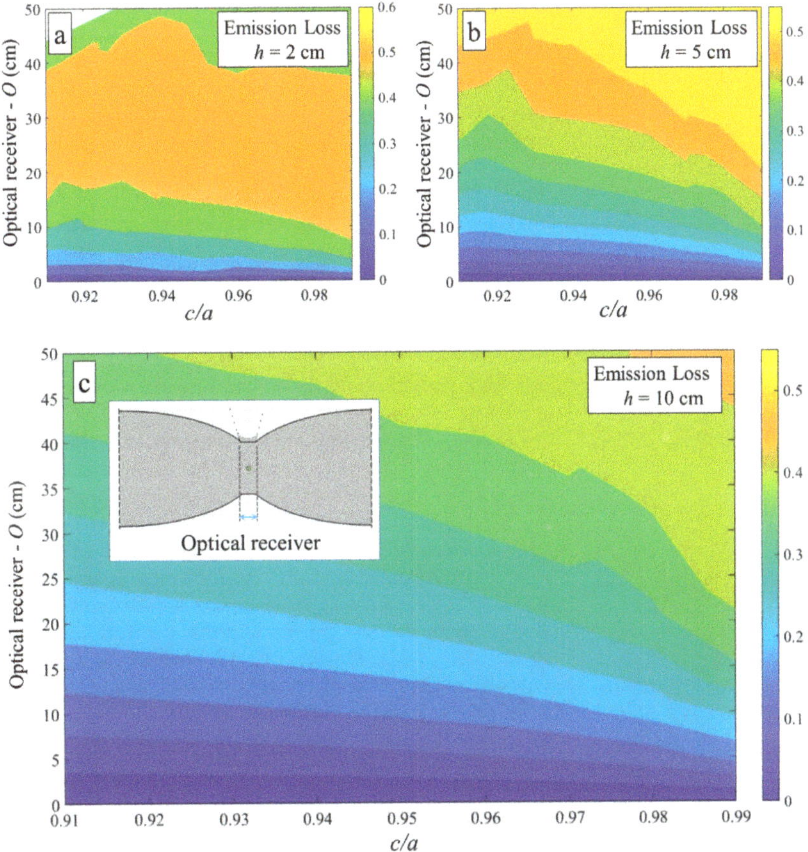

Figure 9. Emission losses as a function of c/a and the width of the optical receiver port (O) for (**a**) $h = 2$ cm, (**b**) $h = 5$ cm and (**c**) $h = 10$ cm.

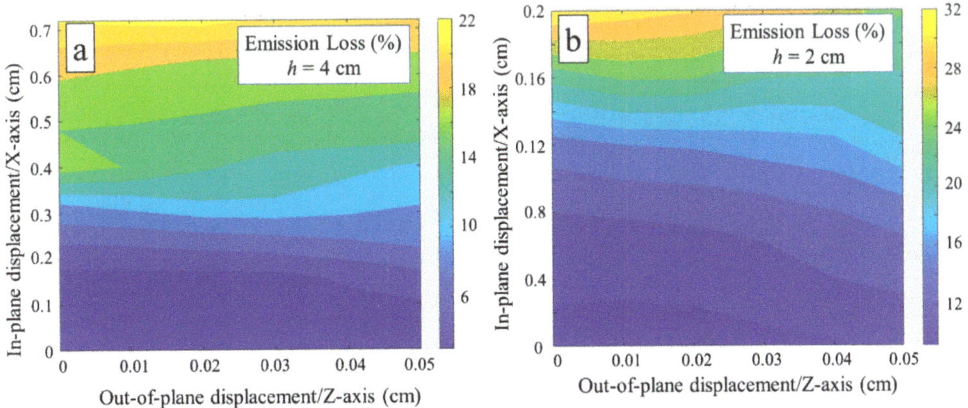

Figure 10. The percentage of emission loss as a function of in-plane and out-of-plane displacement due to the volume distribution of luminescent dye along the focal line for $c = 5$ cm, $O = 1$ mm, and (**a**) $h = 4$ and (**b**) $h = 2$ cm.

As an example, based on the assumptions and analysis regarding dye concentration and absorption that we reported in a previous work [49], it was concluded that the required volume of Lumogen R305 dye in the EASSS with parameter values of $c/a = 0.91$, $c = 1.25$ cm and $O = 0.1$ cm and an area of 10×10 cm^2 to absorb all the incident light over the spectral range of 575~620 nm, requires ~1.57×10^{-3} mm^3 for each unit cell. This required volume creates negligible focal line-offset losses (less than ~1%).

4.5. Performance Comparison: EASSS vs. Conventional LSC Panels

In this section we compare a conventional LSC and an EASSS panel with parameter values of $c/a = 0.91$, $c = 1.25$ cm and $O = 0.1$ cm. The area of both panels is assumed to be 10 cm $\times$ 10 cm, and the height ($h = 2b$) of both panels is assumed to be identical.

The fraction of radiation emitted from a point source located at the center of the panel that is incident onto PV cells located at the panel sidewalls is plotted as a function of time in Figure 11 for the conventional LSC and EASSS panels. Two sidewalls have PV cells and the other two sidewalls are coated by a specular reflector (mirror). The use of sidewall mirrors is discussed in more details in Ref. [50]. At a steady state, the emission losses for the EASSS and conventional LSC panels are 7.4 and 25.8%, respectively, resulting in a 24.8% improvement in emission efficiency. The mean traveling time for photons to propagate from the center of the EASSS and conventional LSC panels to the sidewalls is $T_{m1} = 0.66$ ns and $T_{m2} = 1.45$ ns, respectively. The difference between the mean propagation time for the two panels is related to the differences in transmission efficiency, $\eta_{Transmission}$ (here, transmission efficiency refers to losses that occur for light propagating within the host of the LSC toward its sidewalls). For example, assuming the same intensity of incident light on the EASSS and conventional LSC panels, and that the host material has an index of refraction and mean absorption coefficient of $n = 1.49$ and $\alpha_{mean} = 3 \times 10^{-3}$ cm^{-1}, respectively [48], the Optical Density (OD) can be determined for a light emitted from the center of the EASSS or a planar LSC using the following equation:

$$OD = \log_{10}\left(\frac{I_0}{I}\right) \tag{2}$$

where $I_0 =$ the intensity of light at the center of the LSC and $I =$ the intensity after the light has propagated to the sidewalls of the LSC. The ratio of the optical density for the EASSS and planar LSCs is $OD_{EASSS}/OD_{LSC} = 0.45$, which implies that the transmission losses through the EASSS panel are significantly less than that for the conventional LSC panel.

The total efficiency of the EASSS can be calculated using the concentration factor, C, which is the intensity of the radiation incident onto the rectangular PV cells embedded on two sidewalls, Φ_2, divided by the intensity of the radiation incident onto the Petzvel lens concentrator, Φ_1: $C = \Phi_2/\Phi_1$. In other words, the concentration factor can be re-written as $C = G.\eta_{green/non\text{-}PAR\text{-}opt}$, where G is the geometrical gain factor and $\eta_{green/non\text{-}PAR\text{-}opt}$ is the optical efficiency of the EASSS for the green PAR and non-PAR spectrum, which is given as follows:

$$\eta_{green/non\text{-}PAR\text{-}opt} = \eta_{Fresnel} \cdot \eta_{green/non\text{-}PAR\text{-}Abs} \cdot \eta_{PLQY} \cdot \eta_{Stokes} \cdot \eta_{Emission} \cdot \eta_{Roughness} \cdot \eta_{Transmission} \cdot \eta_{self\text{-}abs} \tag{3}$$

where $\eta_{Fresnel}$ depicts the Fresnel law-based efficiency with which incident light is coupled into the EASSS, instead of being reflected from the boundary surfaces. Taking into consideration the two negative Petzvel lenses, incident light will pass through five surfaces prior to entering the EASSS. If the EASSS and two negative Petzvel lenses are made from PMMA ($n = 1.49$), and for normal incidence, then $\eta_{Fresnel} \cong 0.96^5 \cong 82\%$. $\eta_{green/non\text{-}PAR\text{-}Abs}$ is the efficiency with which the luminophores absorbs the green PAR and non-PAR. Here, we assume that a sufficient amount of luminescent material such as Lumogen Red 305 is embedded in the EASSS (~33 ppm) such that 100% of the green PAR in the spectral region from ~500 to ~600 nm is absorbed [51]. η_{PLQY} represents the photoluminescence quantum yield (PLQY) efficiency described as the ratio between the number of photons emitted and absorbed by the luminescent dye in the green PAR and non-PAR range. It is assumed that $\eta_{PLQY} = 0.95$ for LR 305 luminescent dye [52]. η_{Stokes} shows the energy lost through molecular vibrations and heat generation during the absorption (~580 nm peak) and emission (~620 nm peak) event, which is assumed to be $\eta_{Stokes} = 95\%$ for Lumogen Red 305 [52]. However, in this study, we set $\eta_{Stokes} = 1$ because we are investigating the fraction of incident green PAR photons in the spectral region in the vicinity of 500–600 nm that are absorbed by Lumogen Red 305 and, eventually, directed toward the rectangular PV cells located at the two sidewalls of the EASSS. That is, all the photons incident onto the c-Si PV cell with energy greater than its bandgap, which is ~1.1 eV (λ_{bg} ~1.1 µm), independent of the Stokes shift, contribute to the output current. $\eta_{Emission}$ is the efficiency with which the LSC panels "trap" the light, thus preventing the photons emitted from the luminophores from escaping the panel. As discussed previously with reference to Figure 11, the emission losses for an EASSS with parameters $c/a = 0.91$, $c = 1.25$ and $O = 0.1$, are 7.4%, and, in this case, $\eta_{Emission} = 92.6\%$. In comparison, the $\eta_{Emission}$ for a planar LSC with an index of refraction of 1.49 is ~74%, which is only valid for a perfectly smooth interface. Surface roughness creates parasitic surface losses and reduces the efficiency of total internal reflection and, therefore, an additional factor, $\eta_{Roughness}$, is introduced in Equation (3). These surface scattering losses increase with increasing surface roughness. In practice, surface scattering losses are highly dependent on the quality of fabrication and in an LSC panel can be described using the expression $(1 - D)^N$, where D is the probability that a photon is scattered out of the LSC due to surface roughness each time it is incident onto the LSC surface and N is the number of collisions with the LSC surface [53]. In this study, for the specific case wherein $c/a = 0.91$, $c = 1.25$ cm and $O = 0.1$ cm of the EASSS, it is further assumed that $D = 0.03$, which results in $\eta_{Roughness} = 93\%$. $\eta_{Transmission}$ is the efficiency with which photons generated from the luminescent dye are transported through the LSC without being absorbed by the host material due to the concentrations of attenuating species in the material sample. The transmission efficiency is governed by the following Lambert–Beer law:

$$\frac{I_{out}}{I_{in}} = e^{-\alpha l} \tag{4}$$

where α is the absorption coefficient of the material the panel is comprised of (PMMA in this study). The absorption coefficient is assumed to be $\alpha = 5 \times 10^{-3}$ cm^{-1} for $\lambda = 620$ nm [48] and, therefore, the transmission efficiency is $\eta_{Transmission} = 93.6\%$. $\eta_{self\text{-}abs}$ is the efficiency with which the re-emitted light can be transported through the EASSS panel without being absorbed by other luminophores. Self-absorption is a consequence of the spectral overlap between the emission and absorption spectra of the luminescent dye [54,55]

It should be noted that, self-absorption losses decrease significantly as the Stokes shift increases. However, the total probability of the self-absorption for a photon emitted from a dye is a function of L, the total transport optical path length through the whole medium occupied with luminescent dyes. $P_0(L)$ is the overall probability that an emitted photon will reach the edge of the waveguide without undergoing self-absorption. $P_0(L)$ can be used to calculate the self-absorption losses [55,56] as follows:

$$\eta_{self\text{-}abs} = \frac{P_0(L)}{1 - (1 - P_0(L))\eta_{PLQY} \cdot \eta_{Emission}} \tag{5}$$

where η_{PLQY} is the photoluminescent quantum yield and $\eta_{Emission}$ is the transmission (trapping) efficiency. $\eta_{PLQY} = 0.95\%$ and $\eta_{Emission} = 93\%$ in this study. Therefore, we neglect self-absorption for the proposed LSC in this work.

Considering all the loss mechanisms and efficiencies as described by Equation (5), the green PAR optical efficiency and concentration factor for the EASSS panel considered in this section are $\eta_{green/non\text{-}PAR\text{-}opt} = 61.5\%$ and $C = 2.77$, respectively.

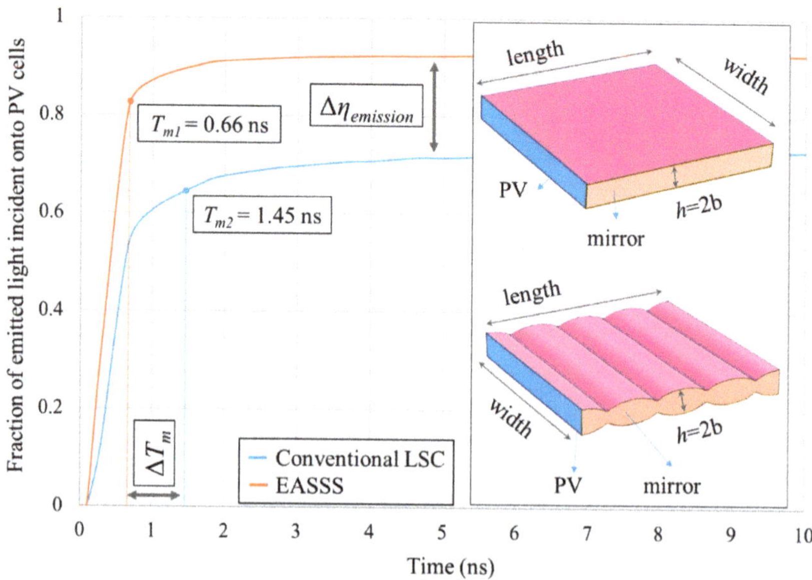

Figure 11. Fraction of photons emitted within the EASSS and conventional LSC panels incident onto PV cells located at the panel sidewalls as a function of time.

It should be mentioned that the self-absorption losses can be the dominant loss mechanism in most cases when luminescent dyes have a lower quantum yield and/or a smaller Stokes shift. For every re-absorption event, the photon energy can be dissipated again through one of the above-mentioned loss channels. Consequently, for an irradiance input energy flux of Φ_{in}, the output flux Φ_{out} decreases after an average re-absorption number of $N - 1$ per photon to [54] the following:

$$\Phi_{out} = \Phi_{in} \cdot \eta_{Fresnel} \cdot \eta_{green/non\text{-}PAR\text{-}Abs} \cdot (\eta_{PLQY} \cdot \eta_{Stokes} \cdot \eta_{Emission} \cdot \eta_{Roughness} \cdot \eta_{Transmission})^N \quad (6)$$

This equation shows that losses increase exponentially as the number of re-absorption events increases. As an example, Figure 12 shows $\eta_{green/non\text{-}PAR\text{-}opt}$ for both a conventional LSC and the EASSS panel with $c/a = 0.91$, $c = 1.25$ cm and $O = 0.1$ cm, and their efficiency ratio (η_{EASSS}/η_{LSC}), as a function of self-absorption events, N. As can be seen, as the number of self-absorption events increases, optical efficiencies exponentially decrease and these decreases occur to a much greater extent for the conventional LSC panel as compared to the EASSS panel.

The mean travelling time, distance and surface scattering events for the photons emitted from the luminophores within the conventional LSC and EASSS panels are provided in Table 1, along with the various efficiency factors and the concentration factor for the two LSC panels.

Table 1. Performance parameters for the EASSS and conventional planar LSC panels.

Solar Concentrator	Mean Photon Traveling Time (ns)	Average Traveling Distance (cm)	Scattering Quantity (Times/Photon)	$\eta_{Emission}$ (%)	$\eta_{Transmission}$ (%)	$\eta_{Roughness}$ (%)	$\eta_{green/non\text{-}PAR\text{-}opt}$ (%)	Concentration Factor
EASSS	0.66	13.29	2.38	92.6	93.6	93	63	2.77
LSC	1.45	29.11	6.92	74	86.45	81	47.2	2.07

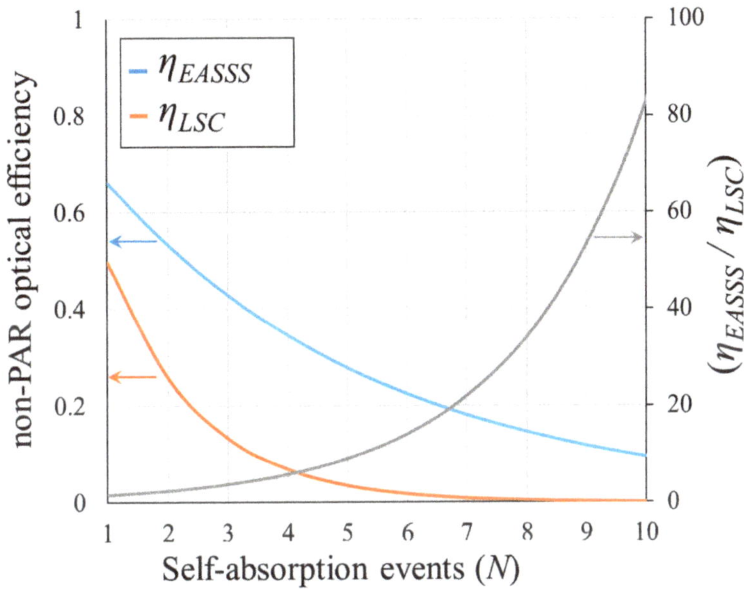

Figure 12. Optical efficiencies for both a conventional LSC and the EASSS panel with $c/a = 0.91$, $c = 1.25$ cm and $O = 0.1$ cm, and their efficiency ratio (η_{EASSS}/η_{LSC}) as a function of self-absorption events (N).

4.6. Thermal Analysis

It is well-known that the quantum efficiency of luminescent dyes degrades as the operating temperature increases [57]. In this study, COMSOL Multiphysics software was utilized to evaluate any potential deleterious thermal effects on the performance of the luminescent dyes throughout the EASSS under the same illumination conditions, as mentioned in the previous sections. In this thermal analysis, the attenuation coefficient and heat transfer coefficient are assumed to be $\alpha_{mean} = 3 \times 10^{-3}$ cm^{-1} and $h = 10$ W/m^2·K, respectively.

The EASSS considered in the previous section, with $c/a = 0.91$, $c = 1.25$ cm and $O = 0.1$ cm, is comprised of a Petzval lens array concentrating solar radiation from a planar area of 100 cm^2 onto four receiver ports that have a width of $O = 1$ mm. The results (explained in the Supplementary Materials, Section B) reveal that the temperature increases by less than 2 degrees throughout the EASSS and the effects of increased temperature on the performance of the EASSS during operation are assumed to be negligible.

4.7. Tandem Elliptic Array LSC-PBR for Combined Power and Algae Production

The inset in Figure 13 illustrates the concept of an EASSS operating in tandem with a PBR. In this example, sunlight is incident onto the topside of a Petzval lens array that directs light to the focal lines of the EASSS. The dye located along these focal lines can absorb the green PAR and non-PAR in the spectral regions from 500~600 nm ($\Delta W_1 = 152$ W/m^2) and 740~1100 nm ($\Delta W_2 = 282$ W/m^2), respectively, as shown as the shaded regions in Figure 13. The light emitted from the dye undergoes TIR and is directed toward the crystalline silicon PV cells located at the sidewalls of the EASSS.

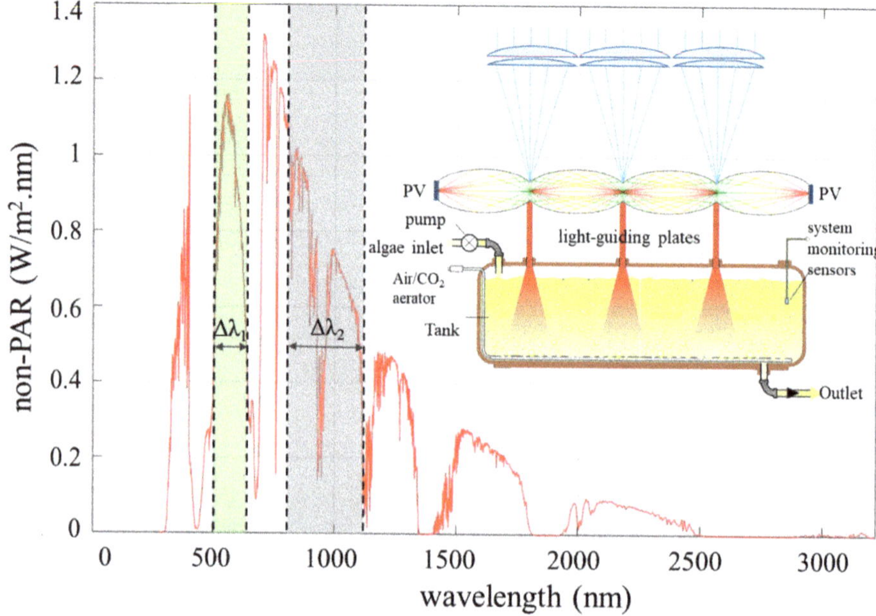

Figure 13. Green PAR and non-PAR from sunlight (ΔW_1 = 500~600 nm and ΔW_2 = 740~1100 nm) can be absorbed in an EASSS in a tandem configuration with a PBR (inset). Blue and red PAR is transmitted through the EASSS and used to support algae cultivation in the PBR.

In this example, it is considered that the green PAR and non-PAR spectral regions from 500~600 nm (ΔW_1 = 152 W/m^2) and 740~1100 nm (ΔW_2 = 282 W/m^2), respectively, can be absorbed in an EASSS and converted to electricity in the c-Si based PV cells at the edges of the EASSS. The Petzval lens array and upper surface area of the PBR shown in Figure 13 are assumed to have an area of A = 10 × 10 cm^2. It is also assumed that the optical efficiency of the EASSS is $\eta_{green/non\text{-}PAR\text{-}opt}$ = 61.5% for both the ΔW_1 and ΔW_2 spectral regions, and the photon flux incident onto the c-Si PV cell at the top side of the EASSS for these spectral regions are $\Phi_{\Delta W1}$ = 4.163 × 10^{16} cm$^{-2}\cdot$s^{-1} and $\Phi_{\Delta W2}$ = 1.263 × 10^{17} cm$^{-2}\cdot$s^{-1}, respectively, which is consistent with receiving solar radiation at an intensity of 1000 W/m^2 (e.g., one sun).

It is assumed the PV cells have an External Quantum Efficiency (EQE) of $EQE^*_{\Delta W*1} \cong 100\%$ and $EQE^*_{\Delta W*2} \cong 85\%$, for light that was absorbed by the dye in spectral regions ΔW_1 and ΔW_2, and reemitted in spectral regions ΔW^*_1 and ΔW^*_2, respectively [58–60]. Under these assumptions, the short circuit current (I_{SC}), open circuit voltage (V_{OC}) and the maximum output power (P_{max}) generated by the PV cell are calculated using the following Equations (7)–(9), respectively [61]:

$$I_{sc} = A \cdot q \cdot \eta_{green/non-PAR-opt} \cdot \left(\int_{\Delta W1} \Phi_{\Delta W1}(\lambda) \cdot EQE^*_{\Delta W*1}(\lambda) \cdot d\lambda + \int_{\Delta W2} \Phi_{\Delta W2}(\lambda) \cdot EQE^*_{\Delta W*2}(\lambda) \cdot d\lambda \right) \tag{7}$$

$$V_{oc} = \frac{nkT}{q} \, ln\left(\frac{I_{sc}}{I_0} + 1 \right) \tag{8}$$

$$P_{max} = I_{sc} \cdot V_{oc} \cdot FF \tag{9}$$

where $n = 1$ is the ideality factor, k is Boltzmann's constant, $T = 300$ K is the cell temperature, $I_0 = 10^{-10}$ A is the dark saturation current, q is the charge of an electron and FF is the fill factor that can be expressed as a function of open-circuit voltage (V_{oc}) as follows [61]:

$$FF = \frac{v_{oc} - \ln\left(v_{oc} + 0.72\right)}{v_{oc} + 1} \tag{10}$$

where

$$v_{oc} = V_{oc}\frac{q}{k_B T} \tag{11}$$

is a normalized voltage. Equation (10) is assumed as an appropriate approximation of the FF when $v_{oc} > 10$. Based on Equation (9), out of a total of 4.34 W of green PAR and non-PAR incident in the ΔW_1 and ΔW_2 solar spectral regions, 0.735 W can be converted to electric power for the surface area of 10 cm $\times$ 10 cm, representing a total green/non-PAR to an electric power conversion efficiency of 16.94% (where $I_{sc} = 1.466$ A, $V_{oc} = 0.605$ V and $FF = 0.8285$).

Notably, this analysis was performed for the AM1.5 spectrum, whereas in space applications, sunlight has an AM0 spectrum. For the solar irradiance of AM0, which is assumed to have an intensity of 1356 W/m^2, by considering the same green/non-PAR optical efficiency of $\eta_{green/non\text{-}PAR\text{-}opt} = 61.5\%$, the conversion efficiency becomes 17.6% ($I_{sc} = 2.066$ A, $V_{oc} = 0.61$ V, $FF = 0.8295$). It means the EASSS can generate 1.45 W out of the total 5.94 W available in ΔW_1 and ΔW_2 for the surface area of 10 cm $\times$ 10 cm. On the other hand, the amount of electric power generated (and amount of PAR available for algae cultivation) strongly depends on the distance from the sun. For example, the maximum solar irradiance on Mars is ~590 W/m^2, compared to a maximum of ~1050 W/m^2 on Earth's surface, and thus, it is expected that the total power (and light available for algae cultivation) that could be generated in the vicinity of Mars would be about half of that generated on Earth.

It should also be noted that the lenses within the Petzval lens array must track the sun, which will require a tracking system. Furthermore, the fabrication methods and stability of the materials used in the EASSS would have to be tested for duration, temperature, vacuum conditions and radiation exposure. In this context, the Petzval lens array could provide shielding for the EASSS. Additionally, with regard to comparing the weights of the EASSS and standard LSC, for a case sample of $c = 5$ cm and $b = 2$ cm, the EASSS is 13.2% lighter. The density of the host material (PMMA) is $\rho = 1190$ Kg/m^3. Detailed analysis about this weight comparison is provided in the Supplementary Materials, Section C.

5. Conclusions

In this work, a new LSC in the form of an elliptic array solar spectrum splitter (EASSS) is presented. Numerical simulations were carried out to compare the performance of the elliptic array LSC to that of a conventional planar LSC. A distinct advantage of the elliptic array LSC is that it can be designed to achieve TIR over a broad range of emission angles, which drastically reduces emission losses. Furthermore, in comparison to a planar LSC, the number of surface scattering events for photons propagating within the EASSS LSC is significantly less than that of a planar LSC, thereby reducing scattering losses. Additionally, the path length traversed by the light travelling from the center to the edge of an EASSS is substantially less than the path length for light moving from the center of a planar LSC to its edge. The shorter path length for light in the EASSS reduces absorption losses, wherein a portion of light emitted from the dye is absorbed in the LSC host material as it travels toward the PV cells located at the edges. Moreover, considering the combined effects of emission, transmission and surface scattering losses, numerical results show the optical efficiency of the elliptic array LSC is 63%, whereas, in comparison, the optical efficiency for conventional planar LSC of the same size is 47.2%. It should be noted that an array of Petzval lenses that track the sun to focus light onto focal lines within the EASSS is required, which increases system cost and complexity. Nevertheless, the EASSS can provide the dual function of partitioning solar irradiance into blue and red PAR and green PAR and non-PAR components to simultaneously provide for algae growth and power generation. The ability to optimally use the broadband solar irradiance for these purposes has potential applications for extended duration space missions wherein power generation and regenerative environmental control and life support systems that include algae growth are of high value. In this context, the results from numerical simulations show that elliptic array luminescent solar concentrators can convert green PAR and non-PAR to electric power with a conversion efficiency of ~17% when the solar irradiance is AM1.5 and 17.6% for AM0 (solar irradiance just outside the atmosphere) while transmitting the remainder of the PAR and non-PAR to an underlying photobioreactor to support algae cultivation.

Supplementary Materials: The following are available online at https://www.mdpi.com/article/10.3390/en14175229/s1, Figure S1: (a) Cross-section of an ellipsoid as a closed loop curve and; (b) the geometrical parameters for an ellipse. A light ray is emitted from f_1 at $t = 0$ ns. The trajectory of this emitted ray changes color in accordance with the colored vertical bar on the right, and after 25 ns the light ray has reached f_2, Figure S2: Temperature distribution within an EASSS unit cell for an incoming solar irradiance of 1000 W/m^2 ($h = 2$ cm and $c = 5$ cm), Figure S3: Temperature profile for the case study ($h = 2$ cm and $c = 5$ cm) of the EASSS along the (a) Top surface of the EASSS shown in the inset with blue line, and (b) Horizontal middle line shown in the inset with red line. Table S1: Mechanical and optical properties of the EASSS for thermal analysis.

Author Contributions: Conceptualization, N.T. and P.G.O.; methodology, N.T. and P.G.O.; software, N.T.; validation, N.T.; formal analysis, N.T.; investigation, N.T.; resources, P.G.O.; data curation, N.T.; writing—original draft preparation, N.T.; writing—review and editing, N.T. and P.G.O.; visualization, N.T.; supervision, P.G.O.; project administration, P.G.O.; funding acquisition, P.G.O. All authors have read and agreed to the published version of the manuscript.

Funding: This research was funded by the Natural Sciences and Engineering Research Council of Canada (NSERC, RGPIN-2017-05987).

Institutional Review Board Statement: Not applicable.

Informed Consent Statement: Not applicable.

Data Availability Statement: Not applicable.

Conflicts of Interest: The authors declare no conflict of interest.

References

1. International Space Exploration Coordination Group. *The Global Exploration Roadmap*; ISECG: Pasadena, CA, USA, 2018.
2. NASA. *NASA's Journey to Mars: Pioneering Next Steps in Space Exploration*; Government Printing Office: Washington, DC, USA, 2015.
3. ESA. *Exploring Together, ESA Space Exploration Strategy*; ESA: Noordwijk, The Netherlands, 2015.
4. Helisch, H.; Keppler, J.; Detrell, G.; Belz, S.; Ewald, R.; Fasoulas, S.; Heyer, A.G. High density long-term cultivation of chlorella vulgaris SAG 211-12 in a novel microgravity-capable membrane raceway photobioreactor for future bioregenerative life support in SPACE. *Life Sci. Space Res.* **2020**, *24*, 91–107. [CrossRef]
5. Revellame, E.D.; Aguda, R.; Chistoserdov, A.; Fortela, D.L.; Hernandez, R.A.; Zappi, M.E. Microalgae cultivation for space exploration: Assessing the potential for a new generation of waste to human life-support system for long duration space travel and planetary human habitation. *Algal Res.* **2021**, *55*, 102258. [CrossRef]
6. Hendrickx, L.; De Wever, H.; Hermans, V.; Mastroleo, F.; Morin, N.; Wilmotte, A.; Janssen, P.; Mergeay, M. Microbial ecology of the closed artificial ecosystem MELiSSA (Micro-Ecological Life Support System Alternative): Reinventing and compartmentizing the Earth's food and oxygen regeneration system for long-haul space exploration missions. *Res. Microbil.* **2006**, *157*, 77–86. [CrossRef]
7. Helisch, H.; Belz, S.; Keppler, J.; Detrell, G.; Henn, N.; Fasoulas, S.; Ewald, R.; Angerer, O. Non-axenic microalgae cultivation in space—Challenges for the membrane μgPBR of the ISS experiment PBR@LSR. In Proceedings of the 48th International Conference on Environmental Systems, Alburquerque, NM, USA, 8–12 July 2018.
8. Sanders, M.H.; Sedwick, R.J. Thermal design and performance of a LSC for space power generation. *J. Spacecr. Rocket.* **2019**, *56*, 1831–1837. [CrossRef]
9. Needell, D.R.; Bauser, H.; Phelan, M.; Bukowsky, C.R.; Ilic, O.; Kelzenberg, M.D.; Atwater, H.A. Ultralight Luminescent Solar Concentrators for Space Solar Power Systems. In Proceedings of the 2019 IEEE 46th Photovoltaic Specialists Conference (PVSC), Chicago, IL, USA, 16–21 June 2019; pp. 2798–2801.

10. Mohsenpour, S.F.; Richards, B.S.; Willoughby, N. Spectral conversion of light for enhanced microalgae growth rates and photosynthetic pigment production. *Bioresour. Technol.* **2012**, *125*, 75–81. [CrossRef]
11. Detweiler, A.M.; Mioni, C.E.; Hellier, K.L.; Allen, J.J.; Carter, S.A.; Bebout, B.M.; Fleming, E.E.; Corrado, C.; Prufert-Bebout, L.E. Evaluation of wavelength selective photovoltaic panels on microalgae growth and photosynthetic efficiency. *Algal Res.* **2015**, *9*, 170–177. [CrossRef]
12. Lamnatou, C.; Chemisana, D. Solar radiation manipulations and their role in greenhouse claddings: Fluorescent solar concentrators, photoselective and other materials. *Renew. Sustain. Energy Rev.* **2013**, *27*, 175–190. [CrossRef]
13. Corrado, C.; Leow, S.W.; Osborn, M.; Chan, E.; Balaban, B.; Carter, S.A. Optimization of gain and energy conversion efficiency using front-facing photovoltaic cell luminescent solar concentrator design. *Sol. Energy Mater. Sol. Cells* **2013**, *111*, 74–81. [CrossRef]
14. Wondraczek, L.; Batentschuk, M.; Schmidt, M.A.; Borchardt, R.; Scheiner, S.; Seemann, B.; Schweizer, P.; Brabec, C.J. Solar spectral conversion for improving the photosynthetic activity in algae reactors. *Nat. Commun.* **2013**, *4*, 2047. [CrossRef]
15. Ooms, M.D.; Dinh, C.T.; Sargent, E.H.; Sinton, D. Photon management for augmented photosynthesis. *Nat. Commun.* **2016**, *7*, 12699. [CrossRef]
16. Yahui, S.; Huang, Y.; Liao, Q.; Fu, Q.; Zhu, X. Enhancement of microalgae production by embedding hollow light guides to a flat-plate photobioreactor. *Bioresour. Technol.* **2016**, *207*, 31–38.
17. Ooms, M.D.; Jeyaram, Y.; Sinton, D. Wavelength-selective plasmonics for enhanced cultivation of microalgae. *Appl. Phys. Lett.* **2015**, *106*, 063902. [CrossRef]
18. Daigle, Q.; O'Brien, P.G. Heat generated using Luminescent Solar Concentrators for Building Energy Applications. *Energies* **2020**, *13*, 5574. [CrossRef]
19. Papakonstantinou, I.; Portnoi, M.; Debije, M.G. The Hidden Potential of Luminescent Solar Concentrators. *Adv. Energy Mater.* **2020**, *11*, 2002883. [CrossRef]
20. Arp, T.B.; Barlas, Y.; Aji, V.; Gabor, N.M. Natural Regulation of Energy Flow in a Green Quantum Photocell. *Nano Lett.* **2016**, *16*, 7461–7466. [CrossRef] [PubMed]
21. Whitmarsh, J. The photosynthetic process. In *Concepts in Photobiology: Photosynthesis and Photomorphogenesis*; Singhal, G.S., Renger, G., Sopory, S.K., Irrgang, K.-D., Govindjee, Eds.; Narosa Publishers and Kluwer Academic: New Delhi, India, 1999; pp. 11–51.
22. Wang, X.; Zhiming, M.W. (Eds.) *High Efficiency Solar Cells: Physics, Materials, and Devices*; Springer Series in Materials Science; Springer Science & Business Media: Berlin/Heidelberg, Germany, 2014.
23. Debije, M.G.; Verbunt, P.P. Thirty years of luminescent solar concentrator research: Solar energy for the built environment. *Adv. Energy Mater.* **2021**, *2*, 12–35. [CrossRef]
24. Correia, S.F.; de Zea Bermudez, V.; Ribeiro, S.J.L.; André, P.S.; Ferreira, R.A.S.; Carlos, L.D. Luminescent solar concentrators: Challenges for lanthanide-based organic–inorganic hybrid materials. *J. Mater. Chem. A* **2014**, *2*, 5580–5596. [CrossRef]
25. Meinardi, F.; Colombo, A.; Velizhanin, K.A.; Simonutti, R.; Lorenzon, M.; Beverina, L.; Viswanatha, R.; Klimov, V.I.; Brovelli, S. Large-area luminescent solar concentrators based on 'Stokes-shift-engineered' nanocrystals in a mass-polymerized PMMA matrix. *Nat. Photonics* **2014**, *8*, 392–399. [CrossRef]
26. Buffa, M.; Debije, M.G. Dye-doped polysiloxane rubbers for luminescent solar concentrator systems. In *High-Efficiency Solar Cells*; Springer: Cham, Switzerland, 2014; pp. 247–266.
27. Zhao, Y.; Meek, G.A.; Levine, B.G.; Lunt, R.R. Near-Infrared Harvesting Transparent Luminescent Solar Concentrators. *Adv. Opt. Mater.* **2014**, *2*, 606–611. [CrossRef]
28. Debije, M.G.; Teunissen, J.P.; Kastelijn, M.J.; Verbunt, P.P.C.; Bastiaansen, C.W.M. The effect of a scattering layer on the edge output of a luminescent solar concentrator. *Sol. Energy Mater. Sol. Cells* **2009**, *93*, 1345–1350. [CrossRef]
29. Cambié, D.; Zhao, F.; Hessel, V.; Debije, M.G.; Noël, T. Every photon counts: Understanding and optimizing photon paths in luminescent solar concentrator-based photomicroreactors (LSC-PMs). *React. Chem. Eng.* **2017**, *2*, 561–566. [CrossRef]
30. Chou, C.H.; Chuang, J.K.; Chen, F.C. High-performance flexible waveguiding photovoltaics. *Sci. Rep.* **2013**, *3*, 2244. [CrossRef] [PubMed]
31. Toledo, C.; Scognamiglio, A. Agrivoltaic Systems Design and Assessment: A Critical Review, and a Descriptive Model towards a Sustainable Landscape Vision (Three-Dimensional Agrivoltaic Patterns). *Sustainability* **2021**, *12*, 6871. [CrossRef]
32. Corrado, C.; Leow, S.W.; Osborn, M.; Carbone, I.; Hellier, K.; Short, M.; Alers, G.; Carter, S.A. Power generation study of luminescent solar concentrator greenhouse. *J. Renew. Sustain. Energy* **2016**, *8*, 043502. [CrossRef]
33. Loik, M.E.; Carter, S.A.; Alers, G.; Wade, C.E.; Shugar, D.; Corrado, C.; Jokerst, D.; Kitayama, C. Wavelength-selective solar photovoltaic systems: Powering greenhouses for plant growth at the food-energy-water nexus. *Earth's Future* **2017**, *10*, 1044–1053. [CrossRef]
34. Bernardoni, P.; Vincenzi, D.; Mangherini, G.; Boschetti, M.; Andreoli, A.; Gjestila, M.; Samà, C.; Gila, L.; Palmery, S.; Tonezzer, M.; et al. Improved Healthy Growth of Basil Seedlings under LSC Filtered Illumination. In Proceedings of the 37th European Photovoltaic Solar Energy Conference and Exhibition, Munich, Germany, 7–11 September 2020; pp. 1767–1771.
35. Talebzadeh, N.; O'Brien, P.G. Selective Solar Concentrators for Biofuel Production and Photovoltaic Applications. In Proceedings of the International Conference of Energy Harvesting, Storage, and Transfer (EHST'17), Toronto, ON, Canada, 21–23 August 2017.
36. Daigle, Q.; Talebzadeh, N.; O'Brien, P.G.; Rauf, I.A. Spectral Splitting Luminescent Solar Concentrator Panels for Agrivoltaic Applications. In Proceedings of the 3rd International Conference of Energy Harvesting, Storage, and Transfer (EHST'19), Ottawa, ON, Canada, 18–19 June 2019; Volume 3, pp. 132–133.

37. Debije, M.G.; Verbunt, P.P.C.; Rowan, B.C.; Richards, B.S.; Hoeks, T.L. Measured surface loss from luminescent solar concentrator waveguides. *Appl. Opt.* **2008**, *47*, 6763–6768. [CrossRef]
38. Verbunt, P.P.C.; Tsoi, S.; Debije, M.G.; Boer, D.J.; Bastiaansen, C.W.M.; Lin, C.W.; de Boer, D.K.G. Increased efficiency of luminescent solar concentrators after application of organic wavelength selective mirrors. *Opt. Express* **2012**, *20*, A655–A668. [CrossRef] [PubMed]
39. Mulder, C.L.; Reusswig, P.D.; Beyler, A.P.; Kim, H.; Rotschild, C.; Baldo, M.A. Dye alignment in luminescent solar concentrators: II. Horizontal alignment for energy harvesting in linear polarizers. *Opt. Express* **2010**, *18*, A91–A99. [CrossRef]
40. Erickson, D.; Sinton, D.; Psaltis, D. Optofluidics for energy applications. *Nat. Photonics* **2011**, *5*, 583–590. [CrossRef]
41. Hincapie, E.; Stuart, B.J. Design, construction, and validation of an internally lit air-lift photobioreactor for growing algae. *Front. Energy Res.* **2015**, *2*, 65. [CrossRef]
42. Xue, S.; Zhang, Q.; Wu, X.; Yan, C.; Cong, W. A novel photobioreactor structure using optical fibers as inner light source to fulfill flashing light effects of microalgae. *Bioresour. Technol.* **2013**, *138*, 141–147. [CrossRef]
43. Sinton, D. Energy: The microfluidic frontier. *Lab Chip* **2014**, *17*, 3127–3134. [CrossRef]
44. Pierobon, S.C.; Riordon, J.; Nguyen, B.; Sinton, D. Breathable waveguides for combined light and CO_2 delivery to microalgae. *Bioresour. Technol.* **2016**, *209*, 391–396. [CrossRef] [PubMed]
45. Lanzarini-Lopes, M.; Delgado, A.G.; Guo, Y.; Dahlen, P.; Westerhoff, P. Optical fiber-mediated photosynthesis for enhanced subsurface oxygen delivery. *Chemosphere* **2018**, *195*, 742–748. [CrossRef]
46. Sun, Y.; Huang, Y.; Liao, Q.; Xia, A.; Fu, Q.; Zhu, X.; Fu, J. Boosting Nannochloropsis oculata growth and lipid accumulation in a lab-scale open raceway pond characterized by improved light distributions employing built-in planar waveguide modules. *Bioresour. Technol.* **2018**, *249*, 880–889. [CrossRef]
47. Huang, J.; Kang, S.; Wan, M.; Li, Y.; Qu, X.; Feng, F.; Wang, J.; Wang, W.; Shen, G.; Li, W. Numerical and experimental study on the performance of flat-plate photobioreactors with different inner structures for microalgae cultivation. *J. Appl. Phycol.* **2015**, *27*, 49–58. [CrossRef]
48. Meinardi, F.; Bruni, F.; Brovelli, S. Luminescent solar concentrators for building-integrated photovoltaics. *Nat. Rev. Mater.* **2017**, *2*, 17072. [CrossRef]
49. Talebzadeh, N.; Rostami, M.; O'Brien, P.G. Elliptic paraboloid-based solar spectrum splitters for self-powered photobioreactors. *Renew. Energy* **2021**, *163*, 1773–1785. [CrossRef]
50. Bernardoni, P.; Mangherini, G.; Gjestila, M.; Andreoli, A.; Vincenzi, D. Performance Optimization of Luminescent Solar Concentrators under Several Shading Conditions. *Energies* **2021**, *14*, 816. [CrossRef]
51. Krumer, Z.; van Sark, W.G.; Schropp, R.E.; de Mello Donegá, C. Compensation of self-absorption losses in luminescent solar concentrators by increasing luminophore concentration. *Sol. Energy Mater. Sol. Cells* **2017**, *167*, 133–139. [CrossRef]
52. Krumer, Z. Self Absorption in Luminescent Solar Concentrators. Ph.D. Thesis, Utrecht University, Utrecht, The Netherlands, 2014.
53. Soleimani, N.; Knabe, S.; Bauer, G.H.; Markvart, T.; Muskens, O.L. Role of light scattering in the performance of fluorescent solar collectors. *J. Photonics Energy* **2012**, *2*, 021801. [CrossRef]
54. Krumer, Z.; Pera, S.J.; van Dijk-Moes, R.J.A.; Zhao, Y.; de Brouwer, A.F.; Groeneveld, E.; van Sark, W.G.J.H.M.; Schropp, R.E.I.; de Mello Donegá, C. Tackling self-absorption in luminescent solar concentrators with type-II colloidal quantum dots. *Sol. Energy Mater. Sol. Cells* **2013**, *111*, 57–65. [CrossRef]
55. Wilson, L.R.; Richards, B.S. Measurement method for photoluminescent quantum yields of fluorescent organic dyes in polymethyl methacrylate for luminescent solar concentrators. *Appl. Opt.* **2009**, *2*, 212–220. [CrossRef]
56. Sansregret, J.; Drake, J.M.; Thomas, W.R.; Lesiecki, M.L. Light transport in planar luminescent solar concentrators: The role of DCM self-absorption. *Appl. Opt.* **1983**, *22*, 573–577. [CrossRef] [PubMed]
57. Rico, F.J.M.; Jaque, F.; Cussó, F. Thermal damage in luminescent solar concentrators (LSC) for photovoltaic systems. *J. Power Sources* **1981**, *6*, 383–388. [CrossRef]
58. Osterwald, C.R.; Campanelli, T.M.; Moriarty, T.; Emery, K.A.; Williams, R. Temperature-dependent spectral mismatch corrections. *IEEE J. Photovolt.* **2015**, *5*, 1692–1697. [CrossRef]
59. Barugkin, C.; Allen, T.; Chong, T.K.; White, T.P.; Weber, K.J.; Catchpole, K.R. Light trapping efficiency comparison of Si solar cell textures using spectral photoluminescence. *Opt. Express* **2015**, *23*, A391–A400. [CrossRef]
60. Chander, S.; Purohit, A.; Nehra, A.; Nehra, S.P.; Dhaka, M.S. A study on spectral response and external quantum efficiency of mono-crystalline silicon solar cell. *Int. J. Renew. Energy Resour.* **2015**, *5*, 41–44.
61. Kippelen, B.; Bredas, J.L. Organic photovoltaics. *Energy Environ. Sci.* **2009**, *2*, 251–261. [CrossRef]

Article

Effects of Nanofluids in Improving the Efficiency of the Conical Concentrator System

Alsalame Haedr Abdalha Mahmood [1], Muhammad Imtiaz Hussain [2,3] and Gwi-Hyun Lee [4,*]

1   Department of Biosystems Engineering, Kangwon National University, Chuncheon 24341, Korea; salami.hayder@gmail.com
2   Agriculture and Life Sciences Research Institute, Kangwon National University, Chuncheon 24341, Korea; imtiaz@kangwon.ac.kr
3   Green Energy Technology Research Center, Kongju National University, Cheonan 31080, Korea
4   Interdisciplinary Program in Smart Agriculture, College of Agriculture and Life Sciences, Kangwon National University, Chuncheon 24341, Korea
*   Correspondence: ghlee@kangwon.ac.kr; Tel.: +82-033-250-6495

Abstract: Fossil fuels are being depleted, resulting in increasing environmental pollution due to greenhouse gases and, consequently, emerging detrimental environmental problems. Therefore, renewable energy is becoming more important; hence, significant research is in progress to increase efficient uses of solar energy. In this paper, the thermal performance of a conical concentrating system with different heat transfer fluids at varied flow rates was studied. The conical-shaped concentrator reflects the incoming solar radiation onto the absorber surface, which is located at the focal axis, where the collected heat is transported through heating mediums or heat transfer fluids. Distilled water and nanofluids (Al_2O_3, CuO) were used in this study as the heat transfer fluids and were circulated through the absorber and the heat storage tank in a closed loop by a pump to absorb the solar radiation. The efficiency of the conical concentrating system was measured during solar noon hours under a clear sky. The collector efficiency was analyzed at different flow rates of 2, 4, and 6 L/min. The thermal efficiency, calculated using different heat transfer fluids, were 72.5% for Al_2O_3, 65% for CuO, and 62.8% for distilled water. Comparing the thermal efficiency at different flow rates, Al_2O_3 at 6 L/min, CuO at 6 L/min, and distilled water at 4 L/min showed high efficiencies; these results indicate that the Al_2O_3 nanofluid is the better choice for use as a heating medium for practical applications.

Keywords: nanofluid; conical concentrator system; performance comparison; thermal efficiency

Citation: Abdalha Mahmood, A.H.; Hussain, M.I.; Lee, G.-H. Effects of Nanofluids in Improving the Efficiency of the Conical Concentrator System. *Energies* **2022**, *15*, 28. https://doi.org/10.3390/en15010028

Academic Editors: Dawei Liang and Changming Zhao

Received: 26 November 2021
Accepted: 20 December 2021
Published: 21 December 2021

Publisher's Note: MDPI stays neutral with regard to jurisdictional claims in published maps and institutional affiliations.

1. Introduction

Recent progressive development of modern technology continues to increase human energy demand. Referring to the energy consumption for domestic use, the proportion of fossil fuels used, such as oil (44%), coal (29%), natural gas (14%), and nuclear power (11%), being very high, and the contribution of new and renewable energies at only 2% [1]. Accordingly, serious environmental pollution problems are emerging; thus, the need for research and development of new and renewable energy is increasing, leading to increasing investment in this sector worldwide. Moreover, the Korean government has established facilitators for renewable energy and clean technologies, such as the Renewable Energy 3020 Plan [2], the power generation gap support system (Feed-in Tariff—FIT), and the renewable energy portfolio standard (RPS); these are indeed strengthening supports for development and distribution projects.

Among the new and renewable energies, solar energy is considered as a useful energy source in our daily life as it has no environmental pollution and is available in abundance [3]. Available solar energy utilization technologies convert sunlight to direct electricity and heat. In particular, solar heat can be used in various fields and has excellent economic benefits [4,5]. However, due to low energy density, it is difficult to use solar energy continuously depending on the outdoor environment. It is evident that the role of a concentrator is very important in solar thermal systems. Therefore, for the efficient usage of solar energy, various types of solar concentrating systems have been developed, including parabolic trough concentrator (PTC)-type, compound parabolic concentrator (CPC)-type, dish-type, and conical-type systems [6–8]. Among these types, the conical concentrator is easier to manufacture and has lower maintenance costs than the other solar concentrating systems. In addition, conical solar concentrator has the advantage of having a smaller absorbing area compared with the flat plate collectors. Furthermore, compared with flat plate collectors, the conical solar collector has excellent heat collection efficiency, which ranges from 60 to 81% [9]. Therefore, in recent years, an immense amount of research in the development of state-of-the-art solar energy collectors has been carried out in the context of improving heat collection efficiency [10]. For solar concentrating systems, the heat collection performance can also be improved by increasing the light collection rate through applying solar tracking technology.

However, research to improve efficiency through structural improvements in solar thermal systems has recently become minimal and has reached its breaking point. In addition, heat transfer fluids used in solar collectors have been limited to water and air. However, recent developments in nanotechnology have led to the development of nanofluids. Nanofluids refer to fluids (as a base fluid) containing nanoparticles with a size of 100 nm or less. Nanofluids have excellent thermal conductivity [11] and have been applied to various fields, such as air-conditioning systems [12], the cooling of electronic devices, and as the heat medium of heat exchangers [13]. The selection of nanofluids is based purely on their economic viability and excellent thermo–physical properties [14]. Based on reported articles in the literature, it has been concluded that Al_2O_3 and CuO are the most widely used heat transfer fluids in solar heat collecting systems.

Many studies are being conducted to maximize solar energy utilization in concentrating solar collecting systems, but research on the heat medium is limited. Therefore, in the proposed study, the outdoor thermal performance of a conical concentrating system using different nanofluids and conventional fluids was discussed. In this paper, distilled water and nanofluids (Al_2O_3 and CuO) were used as the heating mediums. Due to their excellent thermal stability under high temperature range, nanofluids are considered promising alternative to conventional fluids; moreover, due to high solar flux, concentrating solar collectors are capable of producing high temperatures. A combination of the aforementioned solar collectors and the proposed heating mediums into a single unit could be viewed as a viable solution in the context of maximizing the utilization of solar energy. Therefore, the thermal efficiency of a conical solar concentrator using different nanofluids is analyzed and compared with the most commonly used conventional fluids.

2. Materials and Methods

2.1. Conical Concentrating System Configuration and Method

The proposed conical concentrating system consists of conical concentrator that reflects the sun's light, linearly, onto the absorber, a heat storage tank that stores the solar heat, and a centrifugal pump, which is used for the circulation of the heat transfer fluids or heating mediums. The absorber installed at the focal axis of a conical collector is made of copper. Digital flow meters were used to control the flow rate of the working fluid. The extracted solar heat from the conical concentrating system was stored in the thermal storage tank with the help of the heating medium. A schematic of the conical concentrating system is shown in Figure 1.

The conical concentrating system was mounted on a dual-axis tracking platform, which helps to maximize the available solar energy utilization. The experimental facility was located 37° latitude and 127° longitude.

Distilled water and nanofluids (Al_2O_3, CuO) were used as the heat mediums for the conical concentrating system. During operation, the heating medium, stored in the heat storage tank, passes the flow meter by the circulation pump to absorb heat through the absorber surface. The temperature was measured by installing resistance thermometers (Conax Technologies, New York, NY, USA) in the storage tank and at the inlet and the outlet of the conical concentrating system. The measured temperatures were recorded via data loggers (GL820, GRAPHTEC, Irvine, CA, USA). Insolation and meteorological data were measured using a pyrheliometer (Hukseflux, Delft, The Netherlands) and a weathervane (Wireless Vantage). The flow rate of the heat transfer fluid was controlled by the flow meter (PA-60, KOMETER, Incheon, Korea), and the temperature data was recorded in a unit of 1 min and averaged over 10 min.

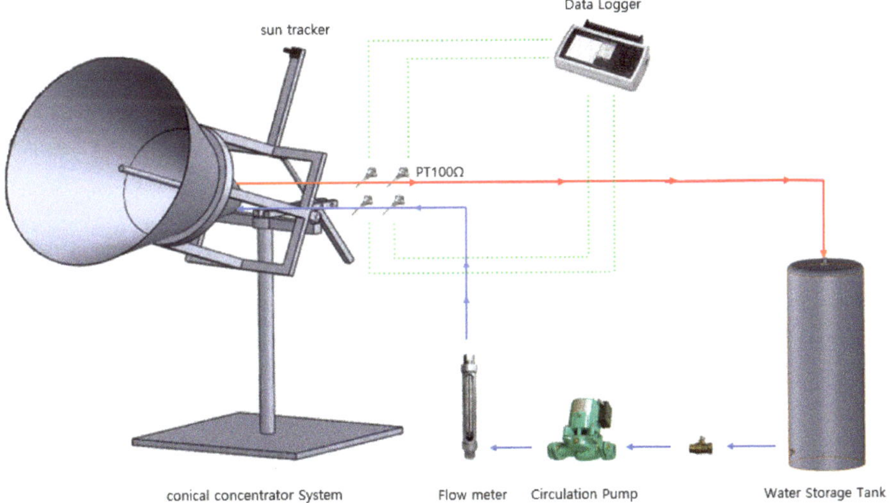

Figure 1. Schematic diagram of the conical concentrator system.

2.2. Nanofluid Manufacturing

In this study, the nanofluid was prepared by a two-step method. Nanofluids were made by dispersing Al_2O_3 and CuO particles, with a size of <50 nm, in distilled water as the base fluid. The nanoparticles used in this study were made by the company AVENTION Co., Ltd. (Incheon, Korea). The thermal conductivity of the nanofluids was analyzed by using different concentrations of surfactant, as suggested by Lee et al. [15]. They used surfactant Cetyltrimethylammonium bromide (CTAB) and Arabic gum (AG) to increase dispersion stability. CTAB was added at 1/10 times, 1 time, and 10 times the critical micelle concentration (CMC), and AG was added at 1/4 times, 1/2 times, and 1 time, based on nanoparticles, because at this point, no CMC concentration was found [15]. Thermal conductivity (W/m °C) was measured by kd2 device. As shown in Table 1, Al_2O_3 and CuO nanofluids had the highest thermal conductivity when 1/10 times of CTAB and 1/2 times of AG, respectively, were added. The nanofluid was prepared by 2 L each; stirring for 30 min using a magnetic stirrer, the prepared mixture is then sonicated for 2 h with an ultrasonic disperser. Besides, stability test for the nanofluids used was conducted at the operating temperature for every 40-day period, and satisfactory stability was found inside the solution with insignificant settling rate. The thermal conductivity of the nanofluids can be calculated by the following correlation [16]:

$$k_{nf} = \frac{\left[\left(k_{np} + 2k_{bf}\right) + 2\phi + \left(k_{np} - k_{bf}\right)\right]}{\left[\left(k_{np} + 2k_{bf}\right) - \phi\left(k_{np} - k_{bf}\right)\right]} \tag{1}$$

where, ϕ is nanoparticles concentration and k_{nf} is the thermal conductivity of the nanofluid. k_{np} and k_{bf} are the thermal conductivities of the nanoparticles and the base fluid, respectively.

Table 1. Thermal conductivity.

Al_2O_3 (0.25%)	Thermal Conductivity (W/m °C)
CTAB 1/10 times	0.851
CTAB 1 time	0.798
CTAB 10 times	0.783
AG 1/4 times	0.805
AG 1/2 times	0.822
AG 1 time	0.826
CuO (0.25%)	**Thermal conductivity (W/m °C)**
CTAB 1/10 times	0.792
CTAB 1 time	0.784
CTAB 10 times	0.771
AG 1/4 times	0.861
AG 1/2 times	0.949
AG 1 time	0.793

2.3. Efficiency Calculation

In this study, the heat collection efficiency of three identical solar concentrating systems was tested at the similar flow rate and operating conditions across the day, where CuO nanofluid, Al_2O_3 nanofluid, and distilled water were used, separately, for each system. The energy performance of the aforementioned solar collectors was carried out at three flow rates—2 L/min, 4 L/min, 6 L/min.

The amount of heat collected (Q) by the absorber is calculated as follows [17]:

$$Q = mC_p(T_o - T_i) \tag{2}$$

where Q and C_p are the flow rate and specific heat, respectively, of the heat transfer fluid. T_i and T_o are the fluid inlet and outlet temperatures, respectively.

The average temperature T_r of the heating medium was calculated using the following Equation (3):

$$T_r = \frac{T_o + T_i}{2} \tag{3}$$

In order to analyze the efficiency of the conical solar concentrator system, the heat collection efficiency (η) was calculated as presented in Equation (4), as follows:

$$\eta = \frac{Q}{Al} \tag{4}$$

where η and l are the thermal efficiency and beam radiation, respectively, and A is the collector area.

3. Results and Discussion

3.1. Efficiency Analysis according to Flow Rate Heat Medium of Conical Concentrating System

3.1.1. Al_2O_3, CuO, and Distilled Water Efficiency Analysis for Flow Rate of 2 L/min

A series of experiments using Al_2O_3, CuO, and distilled water as the working fluids were performed (on 1 November 2018) at a flow rate of 2 L/min under a clear and cloudless sky. To eliminate the error associated with the mass flow rates and the heating medium, three similar systems were tested and compared under the same operating conditions. The outdoor environmental conditions for the experiment are shown in Table 2.

Table 2. The experimental conditions when the flow rate is 2 L/min.

Flow Rate		2 L/min
Solar Radiation (W/m^2)		636.2–860
Wind Speed (m/s)		0.2–1.9
Ambient Temperature (°C)		8.94–17.51
	Al$_2$O$_3$	16.55–66.93
Inlet Temperature (°C)	CuO	16.08–60.95
	distilled water	15.96–54.97

Figure 2 shows the variations of solar radiation and heat collection efficiency over the daily sunshine hours. It was found that the collection efficiency of Al$_2$O$_3$ was the highest. More specifically, the average, highest, and minimum efficiencies using the Al$_2$O$_3$ nanofluid were 67.8%, 73%, and 54%, respectively; whereas, the average, highest, and minimum efficiencies using the CuO nanofluid were found to be 61.4%, 64%, and 53%, respectively. Furthermore, using distilled water, the average, highest, and minimum efficiencies were 58.7%, 62%, and 50%, respectively. The heat collection efficiency decreased with time, and it was judged that convective heat loss increased as the inlet temperature increased. It is observed that the Al$_2$O$_3$ and CuO nanofluids showed better results as heat mediums compared with distilled water due to comparatively higher thermal conductivities.

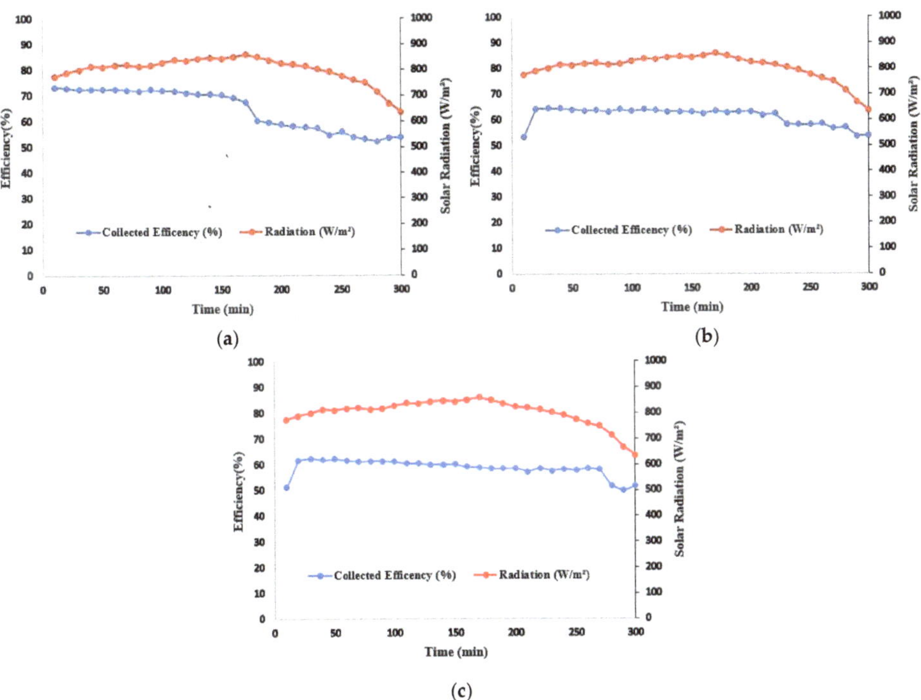

Figure 2. Collected efficiency using (**a**) Al$_2$O$_3$, (**b**) CuO, and (**c**) distilled water at 2 L/min.

3.1.2. Al$_2$O$_3$, CuO, and Distilled Water Efficiency Analysis for Flow Rate 4 L/min

A series of experiments was also conducted using Al$_2$O$_3$, CuO, and distilled water on 2 November 2018, at a flow rate of 4 L/min. The outdoor environmental conditions for the experiment are shown in Table 3.

Table 3. The experimental conditions when the flow rate is 4 L/min.

Flow Rate		4 L/min
Solar Radiation (W/m^2)		740.4–860.2
Wind Speed (m/s)		0–1.9
Ambient Temperature (°C)		10.8–20.7
	Al$_2$O$_3$	22.98–74.67
Inlet Temperature (°C)	CuO	22.7–66.68
	distilled water	22.85–60.9

Figure 3 shows the variations of solar radiation and heat collection efficiency over the daily sunshine hours. The average, highest, and minimum efficiencies using the Al$_2$O$_3$ nanofluid were 65.6%, 70%, and 61%, respectively; whereas, the average, highest, and minimum efficiencies using the CuO nanofluid were found to be 63.8%, 68%, and 52%, respectively. Furthermore, using distilled water, the average, highest, and minimum efficiencies were lower, at 62.8%, 71%, and 55%, respectively. Here, we can note that the Al$_2$O$_3$ and CuO nanofluids showed better results as heat mediums compared with distilled water due to comparatively higher thermal conductivity.

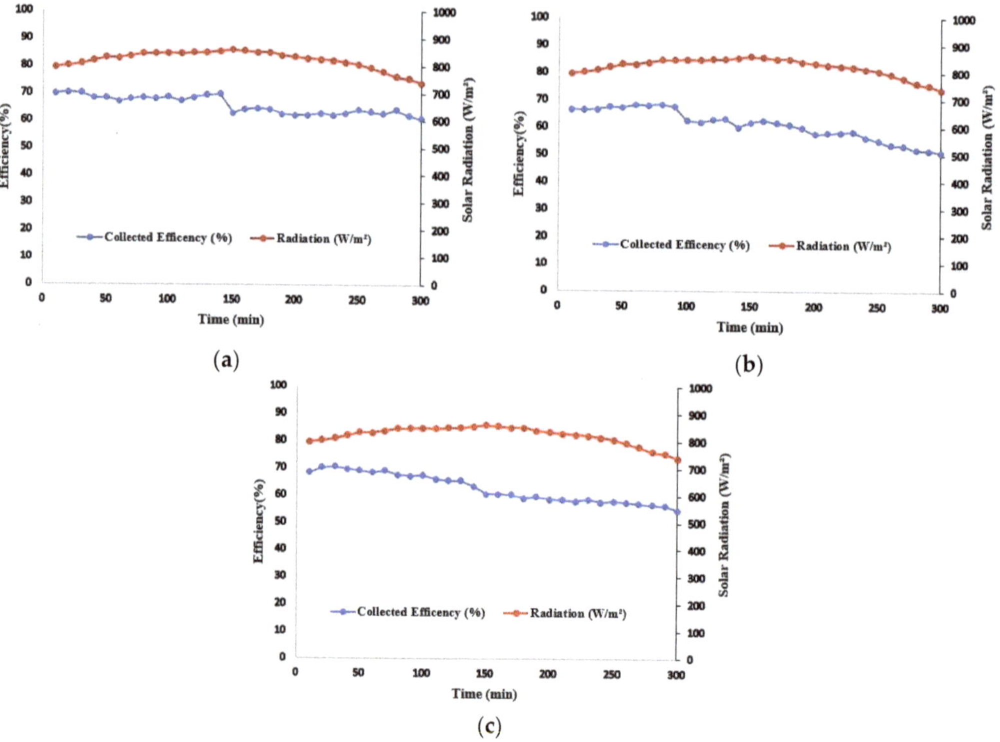

Figure 3. Collected efficiency using (**a**) Al$_2$O$_3$, (**b**) CuO, and (**c**) distilled water at 4 L/min.

3.1.3. Al$_2$O$_3$, CuO, and Distilled Water Efficiency Analysis for Flow Rate 6 L/min

Following a similar experimental procedure, further experiments were conducted on 4 November 2018, using a different flow rate of 6 L/min. The outdoor environmental conditions for the experiment are depicted in Table 4.

Table 4. The experimental conditions and when the flow rate is 6 L/min.

Flow Rate		6 L/min
Solar Radiation (W/m^2)		696.7–869.3
Wind Speed (m/s)		0–2.1
Ambient Temperature (°C)		10.48–22.28
	Al$_2$O$_3$	21.72–79.66
Inlet Temperature (°C)	CuO	21.31–68.8
	Distilled Water	21.62–50.86

Figure 4 shows the variations of solar radiation and heat collection efficiency over the daily sunshine hours. It was found that the collection efficiency of the Al$_2$O$_3$ nanofluid was the highest. More specifically, the average, highest, and minimum efficiencies using the Al$_2$O$_3$ nanofluid were 72.5%, 87%, and 57%, respectively; whereas the average, highest, and minimum efficiencies using the CuO nanofluid were found to be 65%%, 77%, and 42%, respectively. Furthermore, using distilled water, the average, highest, and minimum efficiencies were lower, at 52.2%, 57%, and 46%, respectively.

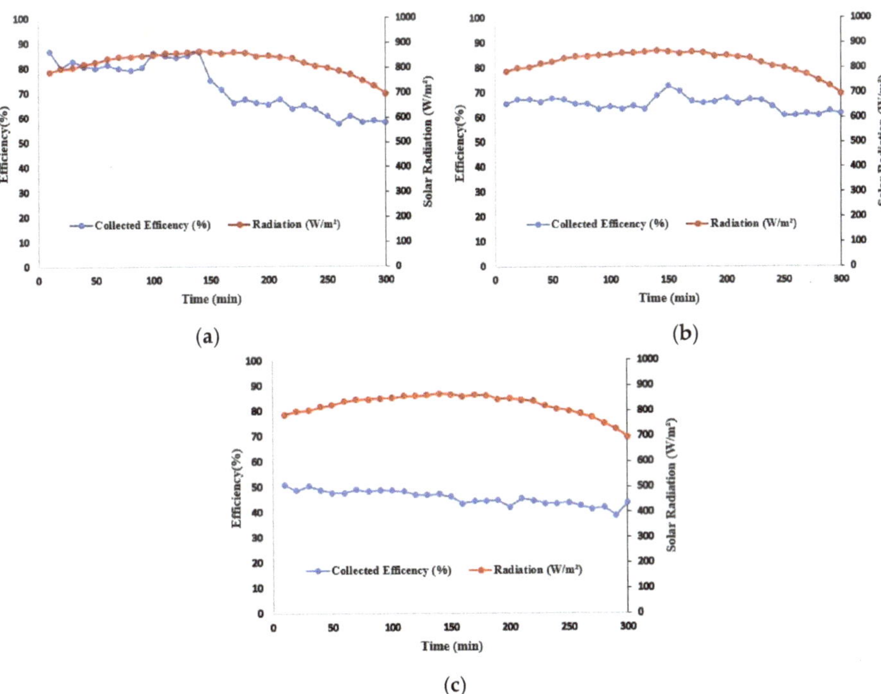

Figure 4. Collected efficiency using (**a**) Al$_2$O$_3$, (**b**) CuO, and (**c**) distilled water at 6 L/min.

The heat collection efficiency decreased with time, and it was judged that convective heat loss increased as the inlet temperature increased. The heat collection efficiency using the Al$_2$O$_3$ nanofluid was the highest. It was observed that the Al$_2$O$_3$ nanofluid absorbs a greater amount of heat than water and CuO under similar ambient conditions; therefore, it is concluded that the relatively high thermal conductivity characteristics of the Al$_2$O$_3$ nanofluid facilitate heat transfer more successfully, resulting in higher efficiency than water and CuO.

3.1.4. Heat Collection Efficiency according to Change of $(T_i - T_a)/I$

Figure 5 shows the results of analyzing the heat collection efficiency according to the change of $(T_i - T_a)/I$. As the temperature difference between the working fluid and the outside air temperature increases, the collection efficiency decreases. The decrease in efficiency is caused by convection heat losses between the absorber surface and the ambient air temperature. It is concluded that the present system had a higher efficiency in comparison with previously published results; nanofluids have shown better results as heating mediums compared with distilled water. In addition, the Al_2O_3 nanofluid was found to be more efficient than CuO nanofluids due to comparatively higher thermal conductivity.

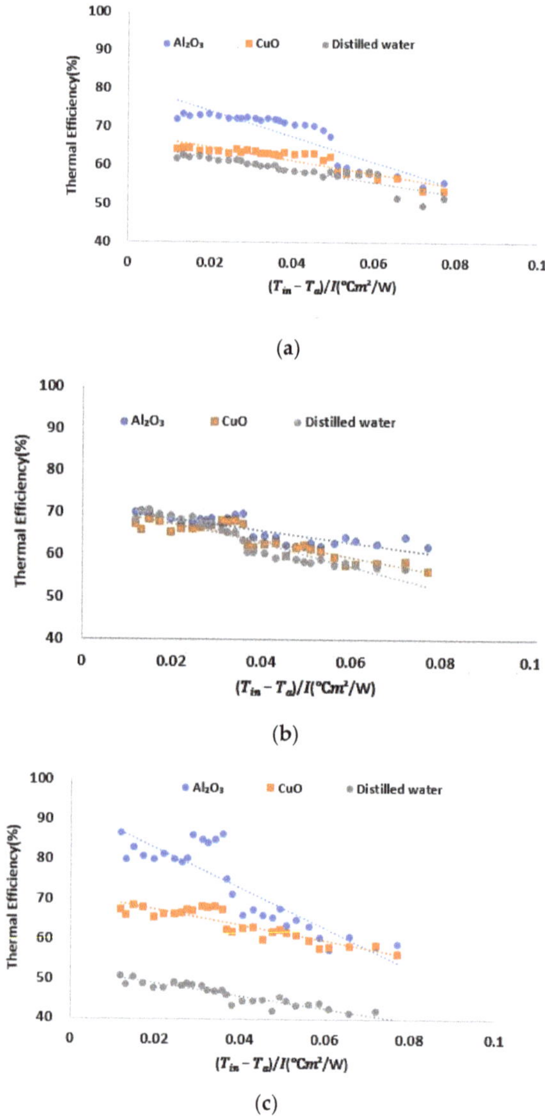

Figure 5. Reduced zero temperature efficiency at (**a**) 2 L/min, (**b**) 4 L/min, and (**c**) 6 L/min.

4. Conclusions

In this study, we analyzed the thermal efficiency of a conical solar collector using nanofluids and conventional fluids. Considering different heat transfer fluids at variable flow rates, the heat collection efficiency for the Al_2O_3 nanofluid at flow rates of 2, 4, and 6 L/min were found to be 65.6%, 67.8%, and 72.5%, respectively; whereas, the CuO nanofluid and the distilled water showed lower efficiencies under similar applied conditions.

Compared with the distilled water, the higher efficiency in the cases of the Al_2O_3 and CuO nanofluids can be explained by their superior thermo–physical properties, which help to extract the extra heat accumulated at the absorber surface. Furthermore, distilled water showed marginal changes in efficiency at all the flow rates from 4 to 6 L/min; therefore, it is clear that the distilled water had the lowest thermal conductivity compared with the nanofluids. Moreover, it was deduced that all the heat accumulated in the absorber was not well recovered by the distilled water, even at high flow rate.

This study was focused on the utilization of nanofluids, especially Al_2O_3 and CuO, as heat mediums for efficient utilization of solar energy in conical solar collector systems. Nanofluids have shown better results as heat mediums as compared with distilled water. In addition, the Al_2O_3 nanofluid was found to be more efficient than CuO due to comparatively higher thermal conductivity. On the basis of the obtained results, the study proposes the practical viability of the nanofluids (especially Al_2O_3) as efficient heat mediums to make maximum use of solar energy as a renewable energy source.

Through this study, it was found that the heat collection efficiency of the conical solar collector was improved using nanofluids as potential heat mediums. However, as the nanofluids circulate continuously through the solar collector, the initial state of dispersion stability is not maintained, and aggregation occurs over time; this may adversely affect the solar collector's performance. Therefore, it is considered necessary to study dispersion stability while circulating nanofluids in the conical concentrating system.

Although nanofluids have higher thermal conductivity than distilled water and their efficiency is high, their heat loss is also high, and it is necessary to study heat loss prevention to improve efficiency.

To increase the absorption rate of available sunlight, painting with Vantablack is recommended, because Vantablack paint is capable of absorbing up to 99.965% of light and might be considered a potential solution. In addition, the addition of a copper coil inside the absorber tube could also help to enlarge the surface area of the absorber, hence maximizing the utilization of the solar energy.

Author Contributions: Conceptualization, A.H.A.M.; methodology, A.H.A.M.; formal analysis, A.H.A.M.; investigation, A.H.A.M.; resources, G.-H.L.; data curation, A.H.A.M. and M.I.H.; writing—original draft preparation, A.H.A.M. and M.I.H.; writing—review and editing, M.I.H. and G.-H.L.; visualization, M.I.H. and G.-H.L.; supervision, G.-H.L.; project administration, G.-H.L.; funding acquisition, G.-H.L. All authors have read and agreed to the published version of the manuscript.

Funding: This work was supported by Korea Institute of Planning and Evaluation for Technology in Food, Agriculture, and Forestry (IPET) through Technology Commercialization Support Program, funded by Ministry of Agriculture, Food, and Rural Affairs (MAFRA) (No. 821048-3).

Institutional Review Board Statement: Not applicable.

Informed Consent Statement: Not applicable.

Data Availability Statement: Data sharing not applicable.

Conflicts of Interest: The authors declare no conflict of interest. The funders had no role in the design of the study; in the collection, analyses, or interpretation of data; in the writing of the manuscript, or in the decision to publish the results.

Nomenclature

Q	concentrated heat (W)
m	mass flow rate (kg/s)
C_p	specific heat (J/kg °C)
T_o	outlet temperature of thermal fluid (°C)
T_i	inlet temperature of thermal fluid (°C)
T_a	ambient temperature (°C)
η	thermal efficiency
I	beam radiation (W/m^2)
A	collector area (m^2)
ϕ	nanoparticles concentration
k_{nf}	the thermal conductivity of the nanofluid (W/m·K)
k_{np}	thermal conductivities of the nanoparticles (W/m·K)
k_{bf}	thermal conductivities of base fluid (W/m·K)

References

1. Petroleum, B. *Renewable Energy—BP Statistical Review of World Energy 2018*; British Petroleum: London, UK, 2018.
2. Kim, J.H.; Kim, S.Y.; Yoo, S.H. Public acceptance of the "Renewable Energy 3020 Plan": Evidence from a contingent valuation study in South Korea. *Sustainability* **2020**, *12*, 3151. [CrossRef]
3. Destek, M.A.; Aslan, A. Disaggregated renewable energy consumption and environmental pollution nexus in G-7 countries. *Renew. Energy* **2020**, *151*, 1298–1306. [CrossRef]
4. Hussain, M.I.; Kim, J.T. Conventional fluid-and nanofluid-based photovoltaic thermal (PV/T) systems: A techno-economic and environmental analysis. *Int. J. Green Energy* **2018**, *15*, 596–604. [CrossRef]
5. Stryi-Hipp, G.; Weiss, W.; Mugnier, D.; Dias, P. *Strategic Research Priorities for Solar Thermal Technology*; European Technology Platform on Renewable Heating and Cooling: Brussels, Belgium, 2012.
6. Zhai, H.; Dai, Y.J.; Wu, J.Y.; Wang, R.Z.; Zhang, L.Y. Experimental investigation and analysis on a concentrating solar collector using linear Fresnel lens. *Energy Convers. Manag.* **2010**, *51*, 48–55. [CrossRef]
7. Hussain, M.I.; Ali, A.; Lee, G.H. Performance and economic analyses of linear and spot Fresnel lens solar collectors used for greenhouse heating in South Korea. *Energy* **2015**, *90*, 1522–1531. [CrossRef]
8. Na, M.S.; Hwang, J.Y.; Hwang, S.G.; Lee, J.H.; Lee, G.H. Design and Performance Analysis of Conical Solar Concentrator. *J. Biosyst. Eng.* **2018**, *43*, 21–29.
9. Joon, Y.H. Experimental and Numerical Analysis for Performance Analysis of Conical Solar Heat Collecting System. Master's Thesis, Kangwon National University, Chuncheon, Korea, 2017.
10. Ding, D.; He, W.; Liu, C. Mathematical Modeling and Optimization of VanadiumTitanium Black Ceramic Solar Collectors. *Energies* **2021**, *14*, 618. [CrossRef]
11. Rashidi, M.M.; Nazari, M.A.; Mahariq, I.; Assad, M.E.H.; Ali, M.E.; Almuzaiqer, R.; Nuhait, A.; Murshid, N. Thermophysical Properties of Hybrid Nanofluids and the Proposed Models: An Updated Comprehensive Study. *Nanomaterials* **2021**, *11*, 3084. [CrossRef] [PubMed]
12. Rahman, S.; Issa, S.; Said, Z.; Assad, M.E.H.; Zadeh, R.; Barani, Y. Performance enhancement of a solar powered air conditioning system using passive techniques and SWCNT/R-407c nano refrigerant. *Case Stud. Therm. Eng.* **2019**, *16*, 100565. [CrossRef]
13. Yoon, Y.H.; Shim, J. A Review on Nano Applications: Fluids and Heat Transfer. *J. Korea Soc. Power Syst. Eng.* **2013**, *17*, 3–12. [CrossRef]
14. Park, K.H.; Lee, J.A.; Kim, H.M. Heat conductivity test and conduction mechanism of nanofluid. *J. Korean Soc. Mar. Eng.* **2008**, *32*, 862–868.
15. Lee, J.H.; Hwang, S.G.; Lee, G.H. Efficiency improvement of a photovoltaic thermal (PVT) system using nanofluids. *Energies* **2019**, *12*, 3063. [CrossRef]
16. Maxwell, J.C. *A Treatise on Electricity and Magnetism*; Clarenton Press: Oxford, UK, 1873; Volume 1, pp. V–XIV.
17. Mahmood Alsalame, H.A.; Lee, J.H.; Lee, G.H. Performance Evaluation of a Photovoltaic Thermal (PVT) system using nanofluids. *Energies* **2021**, *14*, 301. [CrossRef]

Article

Thermo-Economic Performance Evaluation of a Conical Solar Concentrating System Using Coil-Based Absorber

Haedr Abdalha Mahmood Alsalame [1], Muhammad Imtiaz Hussain [2,3], Waseem Amjad [4], Asma Ali [5] and Gwi Hyun Lee [1,*]

1 Department of Interdisciplinary Program in Smart Agriculture, Kangwon National University, Chuncheon 24341, Korea; salami@kangwon.ac.kr
2 Agriculture and Life Sciences Research Institute, Kangwon National University, Chuncheon 24341, Korea; imtiaz@kangwon.ac.kr
3 Green Energy Technology Research Center, Kongju National University, Cheonan 31080, Korea
4 Department of Energy Systems Engineering, University of Agriculture, Faisalabad 38000, Pakistan; waseem_amjad@uaf.edu.pk
5 Department of Agricultural and Resource Economics, Kangwon National University, Chuncheon 24341, Korea; asmaali@kangwon.ac.kr
* Correspondence: ghlee@kangwon.ac.kr; Tel.: +82-033-250-6495

Abstract: Pollution and the increase in greenhouse gas (GHG) emissions have long been linked to the world's increasing need for fossil fuels to generate energy. Every day, the energy consumption is increasing; therefore, it is important to improve technologies that use renewable energy sources. With the abundant availability of sustainable energy, solar power is becoming a necessity. However, solar energy has a low energy density and therefore requires a large installation area, which requires heat collection and heat storage technology. Much research is now being done on the conical solar systems to improve efficiency including calculating an optimal cone angle, finding the best flow ratio and the best absorber design, etc. Therefore, in this study, thermal performance of a conical solar collector (CSC) was assessed with a new design of concentric tube absorber (addition of a coil) and compared to the existing circular tube absorber. It was found that 6 L/min flow rate of heating medium (distilled water and CuO nanofluid) gave lower payback period and higher solar fraction of the system in both cases of absorber tube, i.e., without coil and with coil. However, comparatively, thermal efficiency of CSC with coil-based absorber was almost 10–12% higher than conventional system (without coil) regardless of type of heating medium used.

Keywords: collecting efficiency; conical solar concentrator; performance analysis; solar energy

Citation: Alsalame, H.A.M.; Hussain, M.I.; Amjad, W.; Ali, A.; Lee, G.H. Thermo-Economic Performance Evaluation of a Conical Solar Concentrating System Using Coil-Based Absorber. *Energies* **2022**, *15*, 3369. https://doi.org/10.3390/en15093369

Academic Editors: Dawei Liang and Changming Zhao

Received: 7 April 2022
Accepted: 3 May 2022
Published: 5 May 2022

Publisher's Note: MDPI stays neutral with regard to jurisdictional claims in published maps and institutional affiliations.

1. Introduction

The need for renewable energy is increasing due to increasing industrialization and the over use of fossil fuels, both of which have negative consequences for global climate change. Solar energy is one of the cleanest, most plentiful, and environmentally friendly renewable energy sources. Solar water heaters of various configurations and designs have been utilized for diverse purposes to extract the thermal energy from incoming solar radiation [1]. Recent progressive development of modern technology continues to increase human energy needs. Energy consumption and demand has been increasing worldwide, and researchers are involved in meeting future energy needs [2,3]. The current and expected energy sources are not sustainable such as nuclear power (5%), natural gas (22%), coal/peat (27%), and oil (32%) [1].

The consumption of oil, natural gas, coal and nuclear energy is very high compared to renewable energy, which ultimately leads to pollution problems such as acid rains, ozone layer depletion, and global climate change [4,5].

Renewable energy is a source of sustainable power generation and can potentially minimize pollution problems. The use of renewable energy has increased in recent years, but it is still not widespread. Solar energy is gaining popularity as an option to provide sustainable energy [6]. The most commonly used technologies to harness solar energy are photovoltaic (PV) technology to convert solar energy into electricity and solar thermal collectors to convert sunlight into heat. Solar heat can be used in various fields and has excellent economic advantages [7]. However, due to the low energy density, it is difficult to continuously use solar energy depending on the outdoor environment. The role of a concentrator or reflector is very important in solar concentrating systems. Concentrating solar collectors, which have higher concentration ratios than flat-plate collectors, are of particular interest to researchers. Most cylindrical and circular concentrators have been classified as either reflecting or refractive [6,7]. Various types of concentrators have been developed to date, such as: compound parabolic concentrators (CPC), parabolic trough concentrators (PTC), dish type, and conical solar concentrators [8–12]. Among the above concentration technologies, the conical solar concentrators are easier to manufacture and also possess the advantage of having a small heat loss area compared to other concentrating technologies. In addition, compared to a generally flat plate collector, the conical heat collection unit produces a significantly higher thermal efficiency [13].

Various researchers have investigated the performance of solar thermal collectors [14,15]. Zhai et al. [9], for example, conducted an experimental evaluation of a concentrating solar collector with a linear Fresnel lens, attempting to increase the collector's thermal efficiency over a normal evacuated tube collector system. Kostic and Pavlovic [14] discovered that using solar radiation reflectors at a suitable angle improves the thermal efficiency of a thermal collector. Tao et al. [16] developed the computational and physical models to evaluate the coupled heat transfer performance of a solar dish collector system. To validate the test and model constraints, they examined five different types of solar thermal collector models under steady-state and quasi-dynamic situations [17].

A conical concentrator in a solar concentrating system that reflects the incident radiant flux onto an absorber positioned at the focal axis. Smith's innovative conical surface collector [18] can achieve a high concentration ratio. Countless studies have reported solar concentrating systems for water heating, but only a handful have employed a conical concentrator with a dual-axis tracking system. Some researchers [19] discuss the usage of conical concentrators with one-dimensional tracking systems for air heating. Imtiaz Hussain et al. [20] tested the optical performance of a conical solar collector using stainless steel and aluminum mirror reflectors at different reflector view angles of 35°, 40°, and 45°. Results showed that the maximum thermal of solar collector was achieved at optimal reflector view angle of 45°. In another study [21], they discussed the challenges associated with non-uniformity of the solar flux distribution and its impacts on the temperature distribution in the axial and radial directions on the absorber surface of a conical solar collecting system. However, in the literature there is no example of the use of a conical solar concentrating system with an increase in the absorber area for water heating.

This study focuses on improving the thermal efficiency of the conical solar concentrating system by increasing the absorber surface area. Therefore, a copper coil was placed in the absorber. The proposed conical solar concentrating unit consists of a conical concentrator that collects solar radiation, a heat accumulator (absorber) that converts solar radiation into thermal energy, and a heat storage tank.

The overall aim of this research is to evaluate the efficiency improvement of the conical solar system by inserting a coil inside the concentric tube absorber and also the CuO nanofluid as a heating medium. To achieve these goals, the thermal efficiency of the conical solar collector with the existing absorber (without coil) was compared with the new coil-based absorber under similar operating conditions.

2. Materials and Methods

2.1. Conical Condensing System Configuration and Method

The conical concentrating system consists of a dual-axis tracking system equipped with a sun sensor and a drive shaft to track the sun as shown in Figure 1. During operation, the reflected solar radiations is intercepted via the conical concentrator by the coil-based absorber located at the focal axis of the reflector.

Figure 1. Experimental setup of the conical solar collector system.

Mainly, the design parameters of the conical solar concentrating were adopted from the published study, as reported by Na et al. [12]. The reflector is made of stainless-steel having surface area of 0.785 m^2. The concentration ratio of proposed reflector is 10. High-density energy is converted into heat by double tube absorbers and then transferred to the heating medium. A circulation pump is used to forcibly circulate the heating medium through the coil-based absorber to be heated, and finally, the heated fluid is stored in the heat storage tank. To measure the temperature of the heating medium, two temperature sensors (PT 100 Ω) were installed each at the inlet and outlet of the absorber tube. The amount of direct solar radiation and the outside air temperature were measured using preheliometer and thermocouples, respectively. All devices were connected to a computer via a data acquisition unit. To compare the collection efficiency, two conical condensing systems were built and tested.

2.2. Double Absorption Tube

In this study, a coil-based absorber was built and compared to the reference absorber (without coil). The difference between them occurs in the tubes where the copper coil has been added to increase the heat transfer efficiency by increasing the heat transfer surface area.

2.2.1. Existing Absorber Tube

The reference absorber contains two concentric tubes made of corrosion-resistant copper with high thermal conductivity. The two copper tubes are designed as double tubes for the circulation of the heating medium.

The concentric-tube absorber is attached centrally to the focal axis of the conical concentrator. To increase the absorption rate, the outer surface of the absorber was painted matt black. In addition, rubber insulation is provided at the bottom end to minimize heat loss transferred from the absorber tube. Figure 2 shows the details of the dimension, fluid flow direction, and image of the existing double tube absorber.

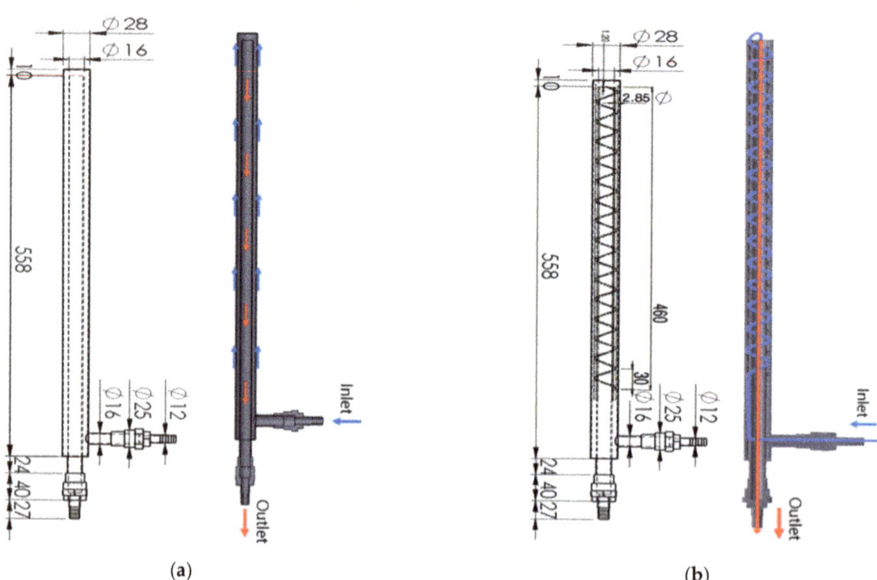

Figure 2. Detail of existing absorber tube with and without coil. (**a**) Absorber tube without coil; (**b**) Absorber tube with coil.

2.2.2. Coil Based Absorber

For the comparison purpose, the dimension details were taken similar to the existing absorber tube. However, the changes occur inside the absorber tube by adding the copper coil between the inner and outer tubes. According to the convection heat transfer as presented in Equation (1), the rate of heat transfer (Q) will get high by enlarging the absorber surface area (A) that is in contact with the working fluid [22,23]. Due to its resistance to corrosion and high thermal conductivity, the coil was made from copper. Details of coil-based absorber are shown in Figure 2.

$$Q = hA\Delta T \tag{1}$$

2.3. Nanofluid Manufacturing

In this study, the CuO nanofluid was prepared by using the two-step process. The specific heat was calculated from the measured values. Maintaining the dispersion stability in the nanofluids is very important to achieve the desired results. The nanofluid manufacturing process is generally divided into a two-stage process and a one-stage process. In the two-stage process, the production of nanoparticles and their dispersion in an available liquid take place separately [24]. This is a method of dispersing the prepared nanoparticles by adding ultrasonic energy to an available base fluid. The advantage of using this method is that it can be mass-produced. In this study, the two-step method was chosen, after adding 0.25 by weight of copper nanoparticles and surfactant with concentration to 400 mL of distilled water as the base liquid, the mixture was stirred for 30 min using a magnetic stirrer (KMC-130SH, VISION, Seoul, Korea). Furthermore, the colloidal solution was sonicated for 1 h and 30 min using an ultrasonic sonicator (KFS-1200N).

The thermal conductivity of the nanofluid was measured using a KD2 Pro Thermal Properties Analyzer (KD2, DRAGON, USA). As shown in Table 1, thermal conductivity of the nanofluids was analyzed by using different concentrations of surfactant as suggested by Abdalha Mahmood et al. [24]. Surfactants CTAB (Cetyltrimethylammonium bromide) and AG (Arabic Gum) were added to improve the stability of the resultant nanofluids. Since the nanofluid temperature rises after ultrasonic dispersion, therefore, for a while the mixture was placed indoors considering the measurement error. Thermal conductivity of the nanofluids was calculated by the following correlation [24]:

$$k_{nf} = \frac{\left[\left(k_{np} + 2k_{bf}\right) + 2\phi + \left(k_{np} - k_{bf}\right)\right]}{\left[\left(k_{np} + 2k_{bf}\right) - \phi\left(k_{np} - k_{bf}\right)\right]} \tag{2}$$

where ϕ represents the nanoparticles concentration and k_{nf}, k_{bf} and k_{np} represent the thermal conductivity of the nanofluid, the base fluid, and of the nanoparticles, respectively.

Table 1. Thermal conductivity of used nanofluid using different concentrations of surfactant.

CuO (0.25%)	Thermal Conductivity (W/m °C)
CTAB 1/10	0.789
CTAB 1	0.781
CTAB 10	0.763
AG 1/2	0.931
AG 1/4	0.864
AG 1	0.788

2.4. Efficiency Calculation

In this work, the thermal efficiencies of the identical conical solar concentrating systems using with and without coil-based absorber were compared. For comparative efficiency analysis, the flow rate is kept constant at 6 L/min, and experiments were carried out during the solar noon hours. Distilled water and CuO nanofluid were used as the heating mediums.

The amount of heat collected (Q) by the absorber tube is calculated as follows:

$$Q = mC_p(T_o - T_i) \tag{3}$$

where, m is mass flow rate of working fluid. C_p and Q are specific heat and the flow rate, respectively, of the heat transfer fluid. T_i and T_o are the fluid inlet and outlet temperatures, respectively.

To analyze the efficiency of the conical solar concentrator system, the heat collection efficiency (η) was calculated as displayed in Equation (4).

$$\eta = \frac{Q}{AS} \tag{4}$$

where, S and η are the input flux radiation and thermal efficiency, respectively, and A is the collector area.

2.5. Economic Analysis

During performance evaluation, the heating cost estimation with both the cases of solar concentrator, i.e., absorber without coil and with coil was done under the same environmental conditions. A cost analysis was performed for both the cases of absorber tube of the solar collector with and without the addition of coil and compared with the cost of heating done using electricity solely. Based on the values of total capital cost, electricity consumption for heating of a room (8 m × 4 m × 3.5 m), discounted payback periods were calculated. Thermal output in both cases was calculated under similar operating and environmental conditions. Table 2 summarized the capital cost and parameters used for cost analysis, i.e., expected useful life of different system components, inflation and interest rates and maintenance cost. The capitals costs were based on the local market.

Table 2. Parameters for cost analysis and capital cost of existing absorber and coil-based absorber using nanofluid as heating medium.

Parameters for Cost Analysis		
Parameters	**Comments**	**Value Range**
Useful life	Collector	30 years
	Absorber	30 years
	Pump	5 years
Mechanical life	Solar tracking system	25–30 years
Inflation rate		2–3%
Interest rate		0.5–2%
Energy cost (electricity)		0.56 USD/kWh
Maintenance cost		1%
Capital cost		
Components	**existing absorber (USD)**	**coil-based absorber (USD)**
Solar tracking system	540	540
Heat storage tank	72	72
Absorber copper tube	21	21
Coil	-	15
Pump	108	108
Supporting structure	36	36
Insulation material	17	13
Total cost	794	805

3. Results and Discussion

The thermal efficiency of the conical concentrating system was compared with and without coil-based absorbers under similar weather conditions. For the reference case, the solar collector performance was investigated with an existing absorber (without coil) and distilled water was used as a heating medium at flow rate 6 L/min. Under clear sky conditions, the time-based data of solar radiation, wind speed, and temperature were collected between 12:06 p.m. and 3:54 p.m. on 5 June 2020. The average values of solar radiation and ambient air temperature were found to be 853.69 W/m^2 and 30.71 °C, respectively, and the wind speed varied in the range of 1.2–1.8 m/s. The solar radiation, water, and air temperatures were measured every minute and averaged over 10 min' intervals. For the existing absorber tube, the inlet and outlet temperatures of working fluids were varied in the range of 32.09–59.06 °C and 33.32–59.94 °C, respectively. While for the coil-based absorber, the inlet and outlet temperatures of working fluids were varied in the range of 31.97–60.91 °C and 33.13–62.09 °C, respectively, as shown in Figure 3.

Similarly, in the case of nanofluid (CuO) as a heating medium, the thermal efficiency of the conical concentrating system was compared with and without coil-based absorbers under similar weather conditions. For the reference case, the solar collector performance was investigated with an existing absorber (without coil) and CuO nanofluid was used as a heating medium at flow rate 6 L/min. Under clear sky conditions, the time-based data of solar radiation, wind speed, and temperature were collected between 10:44 a.m. and 2:32 p.m. on 22 June 2020. the average values of solar radiation and ambient air temperature were found to be 832.6 W/m^2 and 35.76 °C, respectively, and the wind speed varied in the range of 1.3–2 m/s. The solar radiation, water, and air temperatures were measured every minute and averaged over 10 min' interval. For the existing absorber tube, the inlet and outlet temperatures of working fluids were increased from 33.7–68.09 °C and 34.65–69.1 °C, respectively. In the case of a coil-based absorber, the inlet and outlet temperatures of working fluids were increased from 32.9–70.8 °C and 33.4–71.95 °C, respectively, as shown in Figure 4.

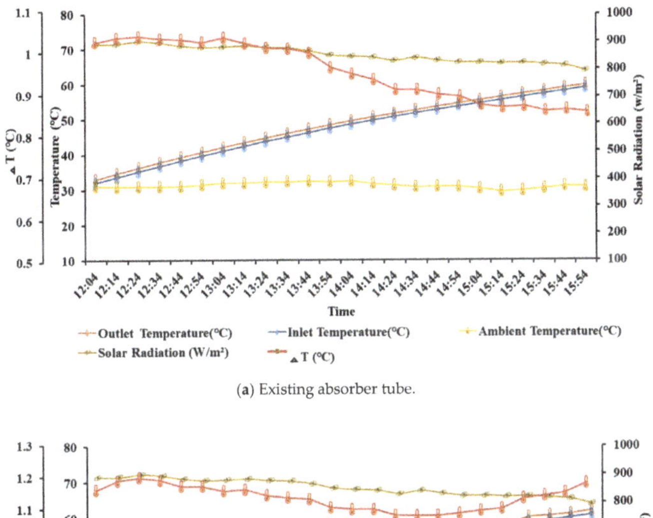

(**a**) Existing absorber tube.

(**b**) Coil-based absorber.

Figure 3. Measured temperature variations and solar radiation in existing absorber tube (**a**) and coil-based absorber (**b**) using distilled water as heating medium (5 June 2020).

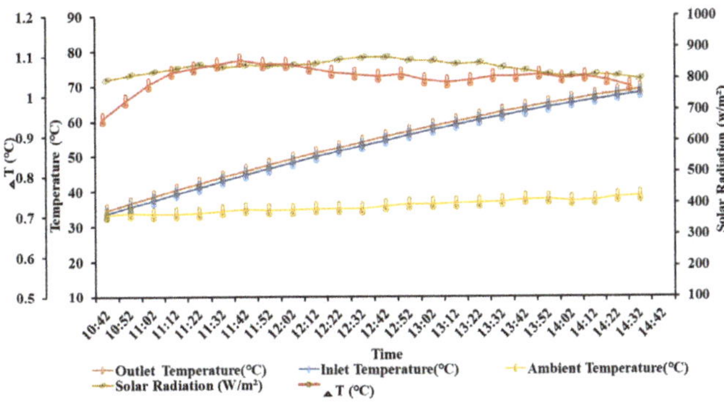

(**a**) Existing absorber tube.

Figure 4. *Cont.*

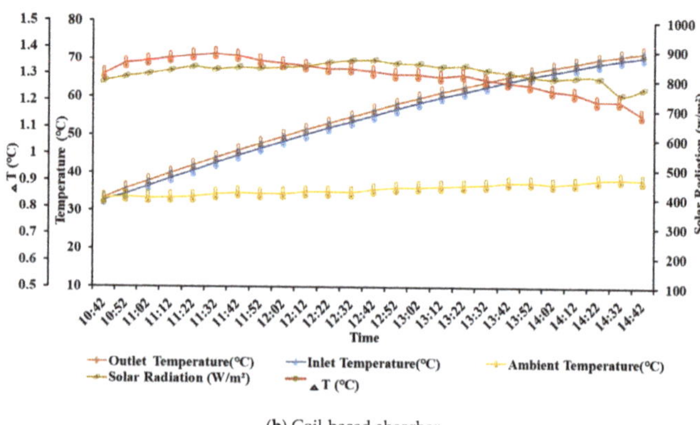

(**b**) Coil-based absorber.

Figure 4. Measured temperature variations and solar radiation in existing absorber tube (**a**) and coil-based absorber (**b**) using CuO-nanofluid as heating medium (22 June 2020).

Based on performance analysis, the maximum, average, and minimum thermal efficiencies of the conical solar collector with an existing absorber (without coil) were found to be 78.4%, 65.1%, and 53%, respectively. In the case of the coil-based absorber, the maximum, average, and minimum thermal efficiencies were found to be 81%, 72.2%, and 58.9%, respectively, as shown in Figure 5 and Table 3.

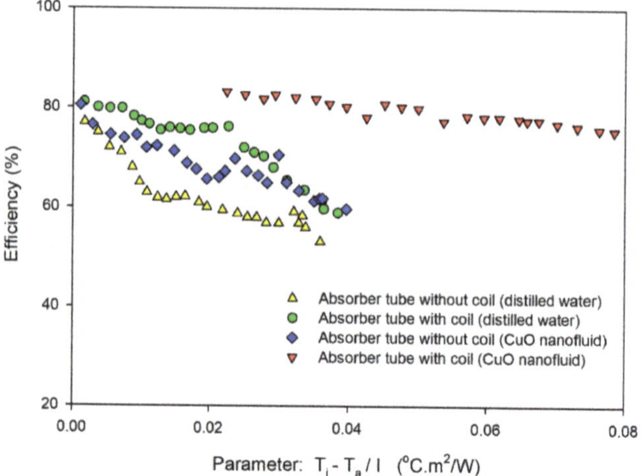

Parameter: $T_i - T_a / I$ (°C.m²/W)

Figure 5. Thermal efficiency analyses of absorber tube without coil and with coil using distilled water and CuO nanofluid as heating medium flowing at rate of 6 L/min.

It can be observed that there is a good correlation between instantaneous efficiency and $(T_i - T_\alpha)/I$ parameter at fixed flow rates of 6 L/min both in cases of existing absorber tube and coil-based absorber. In the case of the existing absorber tube, the average efficiency of the collector increased by 60.1~72.2% with distilled water. The reason for this increase in efficiency at fixed flow rate is due to the elongation of interaction time between the inner absorber surface and the fluid, which is important to an increase in circulating fluid temperature. The high-water temperature leads to more heat loss from the absorber surface than ambient temperature. While using CuO nanofluid in the existing absorber tube, it was found that minimum, average and maximum values of efficiencies were found to be 59.7%, 68.4%, and 80.45%, respectively. Therefore, on an average, using nanofluid, the efficiency of existing absorber tube was found to be 12% more than in the case of using distilled water. On the other side, using the coil-based absorber, the average heat collection efficiency was found to be 68.4% and 79.1% in the case of using distilled water and nanofluid, respectively. Similar to the existing absorber tube case, use of nanofluid gave better efficiency than that of distilled water. Comparing the existing and coil-based absorber, it can be observed that the coil-based absorber showed a 10.7% higher thermal efficiency than that of conventional absorber. It can be explained by the fact that the addition of coil inside the absorber has not only increased the retention time for the exchange of heat between the circulating fluid and the absorber surface, but also enhanced the overall heat transfer area, which ultimately resulted in an enormous thermal gain. This shows that a coil-based absorber using nanofluid as heating medium is good option for better thermal efficiency. It is important to note that the data for each case was taken from different days; therefore, the climatic conditions are different on each day. The inflection at $0.022\,^{\circ}C.m^2/W$ point may be the result of abnormal variations of solar radiation in that span of time.

Table 3. Efficiency of absorber tube without coil and with coil using distilled water and CuO nanofluid as working medium.

Working Medium	Absorber Tube	Average Efficiency (η)	Maximum Efficiency (η)	Minimum Efficiency (η)
Distilled Water	Absorber tube without coil	65.1%	78.4%	53%
	Coil based absorber	72.2%	81%	58.9%
CuO nanofluid	Absorber tube without coil	68.4%	80.45%	59.7%
	Coil based absorber	79.1%	83.1%	75.1%

4. Cost Analysis

Figure 6 shows six months heating cost using electricity and these two cases of solar concentrator. For the months of October and March, the ambient temperature is normally higher than December and January (colder), so cost of electricity is lower and solar concentrator also perform well due to better availability of solar radiations. However, the heating cost due to electricity increased in colder months dye to drop in ambient temperature and existence of more cloudy days. The contribution of the solar absorber with both of its cases (with and without coil) in reducing heating cost is evident. Therefore, economically, it would be the suitable option for space heating, especially in remote areas where the grid supply is limited or nonexistent due to high price of transmission and distribution infrastructure.

For both the cases of absorber tube of the solar collector, payback periods and solar fractions are calculated as a function of flow rate of heating medium as tabulated in Table 4. It can be noted that with the increase of flow rate, the payback periods decreased and solar fractions increased for both the cases and found optimum at flow rate of 6 L per minute. Increasing the flow rate lead to higher heat transfer rate but exceeding the critical limit, i.e., 6 L/min, it started to decrease due to less retention time for the maximum of heat exchange between absorber heat and working medium. For the absorber tube without coil and with coil cases, the minimum payback periods were 14.5 and 11 years, respectively, and the maximum solar fractions were 0.4 and 0.52, respectively, at flow rate of 6 L/min. It can also be noted that for all the cases of flow rates, higher payback periods and lower solar fractions of the absorber tube without coil were calculated than that of respective parameters obtained with the addition of coil. The reasons for the lower payback period with higher solar fraction in the case of absorber with coil was due to the possibility of higher heat transfer rate. With the addition of coil, the heat transfer surface area increased, which facilitated better energy output. Such kinds of collectors can easily be used in other parts of the regions, but the heating efficiency and payback duration would definitely vary depending upon the existing solar radiation and climatic conditions.

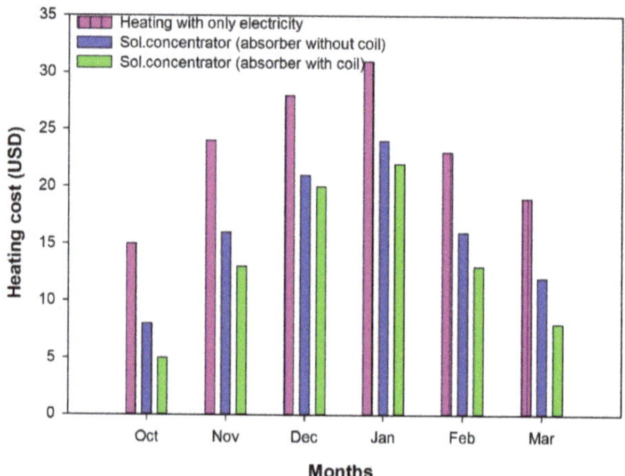

Figure 6. Monthly heating cost using electricity and solar concentrating unit having absorber without coil and with coil.

Table 4. Comparative cost analysis of solar concentrating system having absorber tube with and without a coil under different flow rates of heating medium.

Heating Medium Flow Rate (L/min) in Absorber Tube		Total Capital Cost (USD)		Electricity Savage (USD) Six Months per Year Basis		Payback Period (Years)		Solar Fraction	
Without Coil	With Coil	Without Coil	With Coil	Without Coil	With Coil	Without Coil	With Coil	Without Coil	With Coil
4				43	59	18.5	14	0.30	0.42
5		794	805	45	65	16	12.5	0.35	0.46
6				56	74	14.5	11	0.40	0.52
7				51	68	16	12	0.36	0.48

5. Summary and Conclusions

The efficiency of the conical solar collector was analyzed, comparing the results: (1) between a new coil-based absorber and the existing standard absorber; (2) between CuO nanofluid and distilled water as working fluid. Independently of the heat medium, the new coil-based absorber is more efficient than the existing straight absorber tubes. The CuO nanofluid resulted in a higher efficiency for the whole thermal process because CuO collects heat better than water. The main results are: (1) Using distilled water the heat collection efficiency of the solar collector with coil-based absorber was 12.1% higher than that of the existing straight absorber while for the CuO nanofluid this increase was reduced to 10.7%. The coil-based absorber is clearly more efficient than the existing absorber (without coil). (2) A conical concentrating system with CuO nanofluid showed the highest heat collection efficiency. We believe using the CuO nanofluid as a working fluid is a valuable option when the high-temperature heat is concerned. (3) Since the CuO nanofluid has a higher heat transfer rate than distilled water, the efficiency is high, but heat loss is also massive, so further research on heat loss prevention is necessary to improve efficiency. (4) While the nanofluid circulates through the conical concentrating system, the dispersion stability is not maintained at the initial state, and as time passes, aggregation occurs, so there will be a decrease in efficiency. (5) Changing flow rate of heating medium was found to be a good way to obtain optimum solar fraction and payback period. The absorber tube with coil had a lower payback period and higher solar fraction than the absorber tube without coil at optimum flow rate of 6 L per minute. The contribution of solar absorber to reduce heating expenses in both of its cases (with and without coil) is obvious. Therefore, economically, it is good to use for space heating especially in areas, having poor grid connectivity.

Author Contributions: Conceptualization, H.A.M.A.; methodology, H.A.M.A.; formal analysis, H.A.M.A. and M.I.H.; investigation, H.A.M.A. and M.I.H.; resources, G.H.L.; data curation, H.A.M.A. and M.I.H.; writing—original draft preparation, H.A.M.A. and M.I.H.; writing—review and editing, M.I.H., W.A. and A.A.; visualization, M.I.H. and G.H.L.; supervision, G.H.L.; project administration, G.H.L.; funding acquisition, G.H.L. All authors have read and agreed to the published version of the manuscript.

Funding: This work was supported by the Korea Institute of Planning and Evaluation for Technology in Food, Agriculture, and Forestry (IPET) through the Technology Commercialization Support Program, funded by the Ministry of Agriculture, Food, and Rural Affairs (MAFRA) (No. 821048-3).

Institutional Review Board Statement: Not applicable.

Informed Consent Statement: Not applicable.

Data Availability Statement: Data sharing is not applicable.

Conflicts of Interest: The authors declare no conflict of interest. The funders had no role in the design of the study; in the collection, analyses, or interpretation of data; in the writing of the manuscript, or in the decision to publish the results.

Nomenclature

ϕ	nanoparticles concentration
k_{nf}	the thermal conductivity of the nanofluid (W/m·K)
k_{np}	thermal conductivities of the nanoparticles (W/m·K)
k_{bf}	thermal conductivities of base fluid (W/m·K)
Q	useful energy (W)
m	mass flow rate (kg/s)
C_p	specific heat (J/kg°C)
T_o	the outlet temperature of the thermal fluid (°C)
T_i	the inlet temperature of the thermal fluid (°C)
T_a	ambient temperature (°C)
η	thermal efficiency
I	Beam radiation (W/m^2)
A	collector area (m^2)
h	convective heat transfer coefficient (W/(m^{2}K))
T	Temperature difference between the absorber and ambient air (°C)

References

1. WBA. *Global Bioenergy Statistics 2019*; World Bioenergy Association: Stockholm, Sweden, 2019.
2. Hasanuzzaman, M.; Rahim, N.A.; Saidur, R.; Kazi, S.N. Energy savings and emissions reductions for rewinding and replacement of the industrial motor. *Energy* **2011**, *36*, 233–240. [CrossRef]
3. Hasanuzzaman, M.; Rahim, N.A.; Hosenuzzaman, M.; Saidur, R.; Mahbubul, I.M.; Rashid, M.M. Energy savings in the combustion-based process heating in the industrial sector. *Renew. Sustain. Energy Rev.* **2012**, *16*, 4527–4536. [CrossRef]
4. Dincer, I.; Rosen, M.A. A worldwide perspective on energy, environment, and sustainable development. *Int. J. Energy Res.* **1998**, *22*, 1305–1321. [CrossRef]
5. Colombo, U. Development and the global environment. In *The Energy—Environment Connection*; Hollander, J.M., Ed.; Island Press: Washington, DC, USA, 1992; pp. 3–14.
6. Phuangpornpitak, N.; Kumar, S.P.V. Hybrid systems for rural electrification in Thailand. *Renew. Sustain. Energy Rev.* **2007**, *11*, 1530–1543. [CrossRef]
7. Stryi-Hipp, G.; Weiss, W.; Mugnier, D.; Dias, P. *Strategic Research Priorities for Solar Thermal Technology*; European Technology Platform on Renewable Heating and Cooling: Brussels, Belgium, 2012.
8. Zambolin, E.; Del Col, D. Experimental analysis of the thermal performance of flat plate and evacuated tube solar collectors in stationary standard and daily conditions. *Sol. Energy* **2010**, *84*, 1382–1396. [CrossRef]
9. Zhai, H.; Dai, Y.; Wu, J.; Wang, R.; Zhang, L. Experimental investigation and analysis on a concentrating solar collector using linear Fresnel lens. *Energy Convers. Manag.* **2010**, *51*, 48–55. [CrossRef]

10. Hussain, M.I.; Ali, A.; Lee, G.H. Performance and economic analyses of linear and spot Fresnel lens solar collectors used for greenhouse heating in South Korea. *Energy* **2015**, *90*, 1522–1531. [CrossRef]
11. Zhao, C.; You, S.; Wei, L.; Gao, H.; Yu, W. Theoretical and experimental study of the heat transfer inside a horizontal evacuated tube. *Sol. Energy* **2016**, *132*, 363–372. [CrossRef]
12. Na, M.S.; Hwang, J.Y.; Hwang, S.G.; Lee, J.H.; Lee, G.H. Design and Performance Analysis of Conical Solar Concentrator. *J. Biosyst. Eng.* **2018**, *43*, 21–29.
13. Hwang, J.R. Experimental and Numerical Analysis for Performance Analysis of Conical Solar Heat Collecting System. Master's Thesis, Kangwon National University, Chuncheon, Korea, 2018.
14. Sakhrieh, A.; Ghandoor, A.A.L. Experimental investigation of the performance of five types of solar collectors. *Energy Convers. Manag.* **2013**, *65*, 715–720. [CrossRef]
15. Kumar, R.; Rosen, M.A. Thermal performance of integrated collector storage solar water heater with corrugated absorber surface. *Appl. Eng.* **2010**, *30*, 1764–1768. [CrossRef]
16. Tao, Y.B.; He, Y.L.; Cui, F.Q.; Lin, C.H. Numerical study on coupling phase change heat transfer performance of solar dish collector. *Sol. Energy* **2013**, *90*, 84–93. [CrossRef]
17. Tiago, O.; João, C.M. Testing of solar thermal collectors under transient conditions. *Energy Procedia* **2012**, *30*, 1344–1353.
18. Smith, P.D. Solar Collector with Conical Element. U.S. Patent US 4,030,477, 21 May 1977.
19. Turk, T.I.; Dursun, P.; Cevdet, A. Development and testing of a solar air-heater with conical concentrator. *Renew. Energy* **2004**, *29*, 263–275. [CrossRef]
20. Hussain, M.I.; Lee, G.H.; Kim, J.T. Experimental validation of mathematical models of identical aluminum and stainless steel engineered conical solar collectors. *Renew. Energy* **2017**, *112*, 44–52. [CrossRef]
21. Hussain, M.I.; Lee, G.H. Numerical and experimental heat transfer analyses of a novel concentric tube absorber under non-uniform solar flux condition. *Renew. Energy* **2017**, *103*, 49–57. [CrossRef]
22. Sabharwall, P.; Utgikar, V.; Gunnerson, F. Effect of mass flow rate on the convective heat transfer coefficient: Analysis for constant velocity and constant area case. *Nucl. Technol.* **2009**, *166*, 197–200. [CrossRef]
23. Tomlinson, C.E. *Nuclear Power Plant Thermodynamics, and Heat Transfer*, 1st ed.; Iowa State University Press: Ames, IA, USA, 1989.
24. Abdalha Mahmood, A.H.; Hussain, M.I.; Lee, G.H. Effects of Nanofluids in Improving the Efficiency of the Conical Concentrator System. *Energies* **2022**, *15*, 28. [CrossRef]

Article

Uniform and Non-Uniform Pumping Effect on Ce:Nd:YAG Side-Pumped Solar Laser Output Performance

Cláudia R. Vistas [1], Dawei Liang [1,*], Dário Garcia [1], Miguel Catela [1], Bruno D. Tibúrcio [1], Hugo Costa [1], Emmanuel Guillot [2] and Joana Almeida [1]

[1] Centro de Física e Investigação Tecnológica, Departamento de Física, Faculdade de Ciências e Tecnologia, Universidade NOVA de Lisboa, 2829-516 Caparica, Portugal; c.vistas@fct.unl.pt (C.R.V.); kongming.dario@gmail.com (D.G.); m.catela@campus.fct.unl.pt (M.C.); brunotiburcio78@gmail.com (B.D.T.); hf.costa@campus.fct.unl.pt (H.C.); jla@fct.unl.pt (J.A.)

[2] PROMES-CNRS, 7 rue du Four Solaire, 66120 Font Romeu, France; emmanuel.guillot@promes.cnrs.fr

* Correspondence: dl@fct.unl.pt

Abstract: The Ce:Nd:YAG is a recent active medium in solar-pumped lasers with great potential. This study focuses on the influence of two secondary concentrators: a fused silica aspherical lens and a rectangular fused silica light guide; and consequent pump light distribution on the output performance of a Ce:Nd:YAG side-pumped solar laser. The solar laser head with the aspherical lens concentrated the incident pump light on the central region of the rod, producing the highest continuous-wave 1064 nm solar laser power of 19.6 W from the Ce:Nd:YAG medium. However, the non-uniformity of the absorbed pump profile produced by the aspherical lens led to the rod fracture because of the high thermal load, limiting the maximum laser power. Nevertheless, the solar laser head with the light guide uniformly spread the pump light along the laser rod, minimizing the thermal load issues and producing a maximum laser power of 17.4 W. Despite the slight decrease in laser power, the use of the light guide avoided the laser rod fracture, demonstrating its potential to scale to higher laser power. Therefore, the pumping distribution on the rod may play a fundamental role for Ce:Nd:YAG solar laser systems design.

Keywords: Ce:Nd:YAG; aspherical lens; light guide; solar laser; side-pumped; uniformity

Citation: Vistas, C.R.; Liang, D.; Garcia, D.; Catela, M.; Tibúrcio, B.D.; Costa, H.; Guillot, E.; Almeida, J. Uniform and Non-Uniform Pumping Effect on Ce:Nd:YAG Side-Pumped Solar Laser Output Performance. *Energies* **2022**, *15*, 3577. https://doi.org/10.3390/en15103577

Academic Editor: Tapas Mallick

Received: 20 April 2022
Accepted: 11 May 2022
Published: 13 May 2022

Publisher's Note: MDPI stays neutral with regard to jurisdictional claims in published maps and institutional affiliations.

1. Introduction

The direct conversion of sunlight into laser radiation by solar-pumped lasers is of interest for several laser-based applications such as free-space optical communications, laser propulsion, space-to-Earth wireless power transmission, asteroid deflection, and photovoltaic energy conversion [1–3]. Since early research of solar-pumped laser in the 1960s [4,5], the research of Arashi et al. in 1984 [6], Weskler et al. in 1988 [7], Lando et al. in 2003 [8], Dinh et al. in 2012 [9] and Liang et al. in 2018 [10] have revealed considerable progresses on optical pumping approaches in order to obtain a better solar laser output efficiency. The optical pumping design relies on primary, secondary and tertiary concentrators to achieve enough pumping intensity for the lasing threshold. Primary concentrators, namely parabolic mirrors and Fresnel lenses, collect and concentrate the solar radiation to a focal zone, where a solar laser head is introduced. Secondary concentrators, such as aspheric lenses and light guides, concentrate and distribute the solar light rays from the focal zone of the primary concentrator to the laser-active medium. Tertiary concentrators, such as 2D compound parabolic concentrators (CPCs), 3D CPCs, and V-shaped pump cavities, are also indispensable for high-flux pumping of the laser medium since they can either compress or wrap the concentrated solar radiation from their input aperture to the laser-active medium. Two most widely used methods of solar pumping are end-side-pumping [9–17] and side-pumping [8,18–23]. Although the end-side-pumping approach allows greater laser efficiency, the distribution of pump light is non-uniform, which may lead to serious thermal loading problems [12]. This issue is countered by using the side-pumping method as it allows a more uniform light distribution along the rod axis, spreading the absorbed power within the medium.

The most popular laser gain medium is the Nd:YAG because of its excellent spectroscopic, mechanical, and thermal properties. However, the prominent drawback of the Nd:YAG medium is the poor overlap of its narrow absorption bands with the solar spectrum.

Sensitizers of the Nd^{3+} ion emission, such as Ce^{3+} and Cr^{3+} ions, can be added as co-dopants in the doped YAG host to increase the efficiency of solid-state lasers by absorbing more light from broadband radiation, transferring the excitation energy to the Nd^{3+} ions (discussed below).

The Cr:Nd:YAG active medium has been used in solar-pumped lasers; however, the gain has not been significantly greater than what would be theoretically expected [10,14,24]. The record collection efficiency is 32.5 W/m^2 for an end-side-pumped solar laser using a Cr:Nd:YAG ceramic rod [10]. Nevertheless, this value is only 1.03 and 1.01 times higher than that of a Nd:YAG single-crystal rod [14] and a bonding Nd:YAG/YAG crystal rod [16], respectively.

Just recently, the Ce:Nd:YAG medium was tested in solar lasers by using end-side-pumping [25] and side-pumping [26] configurations. The former produced a lower output power compared with that of the Nd:YAG solar laser under the same solar pump power [25]. This can result from the larger scattering loss within the crystal attributed to the Ce co-dopant, but also from the thermal load on the Ce:Nd:YAG rod by using an end-side-pumping configuration, causing the final fracture of the rod [25]. Therefore, in the most recent study, a side-pumping configuration was used to allow a uniform pump light absorption and homogeneous laser rod heating [26]. This led to a significant increase in the solar laser efficiency, obtaining the record 23.6 W/m^2 collection efficiency for a side-pumping configuration, which was 1.57 times higher than the one obtained with a Nd:YAG solar laser under the same pumping conditions [26].

It is interesting to note that the thermal loading issues are not manifested in the same manner in the Nd:YAG and Ce:Nd:YAG active media. The fracture of a Nd:YAG laser rod already occurred in an end-side-pumped solar laser configuration. However, its incoming solar power was at a much higher level of 2585 W [12] compared with the only 964 W incoming power for the Ce:Nd:YAG rod also in an end-side-pumped solar laser configuration [25]. The main reason behind this difference may rely on the Stokes shift. The Stokes shift (η_s) is the spectral difference between the emission ($\lambda_{Emission}$) and the pump (λ_{Pump}) peak wavelengths. The Nd:YAG pumped by sunlight has a Stokes efficiency of 0.62 ($\eta_s = \lambda_{Pump}/\lambda_{Emission}$ = 660 nm/1064 nm = 0.62 [7]); thereby, 38% of the pump photon energy is dissipated as heat to the crystal. The Ce:Nd:YAG has a lower Stokes efficiency of about 0.43 ($\eta_s = \lambda_{Pump}/\lambda_{Laser}$ = 460 nm/1064 nm = 0.43 [27]); thus, about 57% of the energy of every pump photon transferred from Ce^{3+} to Nd^{3+} ions is lost. This lower Stokes efficiency results in a higher thermal load on the active medium. Therefore, the thermal effects were more pronounced under solar pumping, leading the Ce:Nd:YAG rod to fracture more easily [25] compared with the Nd:YAG rod [12,14].

Since the thermal loading issues play a vital role in the design of Ce:Nd:YAG solar laser systems, it is important to consider the solar pumping distribution along the laser rod on the optimization process of solar laser power scaling. Although the side-pumping configuration may allow a more uniform light distribution along the rod compared with the end-side-pumping configuration it is the secondary concentrator that plays a fundamental role in the distribution and uniformization of the light rays. The two most used secondary concentrators in side-pumped solar lasers are the fused silica aspherical lens [13,14,18,23] and fused silica light guide [19,21,22]. Even though these concentrators have been frequently applied in solar laser prototypes, a comparative study between them to evaluate their performance on solar lasers is still lacking. Since Ce:Nd:YAG is a new and potential active medium in solar-pumped lasers, it is important to study the effect of the pump light distribution on its performance. Therefore, this study is focused on the influence of these secondary concentrators and consequent pump light distribution on the output performance of the Ce:Nd:YAG side-pumped solar laser.

2. Optical Spectra and Energy Transfer from Ce^{3+} to Nd^{3+} Ions in Ce:Nd:YAG Medium

The Ce:Nd:YAG active medium has already demonstrated the potential to increase the efficiency of broadband-pumped lasers, compared with the Nd:YAG crystal [28–32] because of its strong absorption in ultraviolet (UV) and visible spectral regions at broadband light pumping and efficient energy transfer from Ce^{3+} to Nd^{3+}. The solar irradiance spectrum, the Nd:YAG and Ce:Nd:YAG absorption spectra, and the Ce:YAG fluorescence spectrum are represented in Figure 1. The Ce:Nd:YAG absorption spectrum has two broad absorption bands centered at 339 nm and 460 nm, which are characteristic of the Ce^{3+} ion in YAG lattice and other bands that are characteristic of the Nd^{3+} ions. The Ce:YAG fluorescence spectrum has a strong and broad green luminescence centered around 531 nm that overlaps well with two absorption peaks of the Nd^{3+} ion around 530 nm and 589 nm.

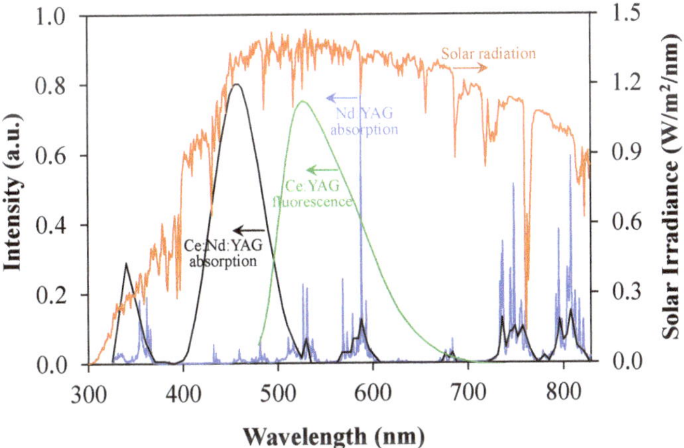

Figure 1. Standard solar emission spectrum (orange line) [33], Ce:YAG fluorescence spectrum (green line) (adapted from [34]), Nd:YAG (blue line) (adapted from [35]), and Ce:Nd:YAG (black line) (adapted from [36]) absorption spectra.

The energy transfer mechanisms in Ce:Nd:YAG material have been investigated in several studies [28–30,36–38]. Radiative energy transfer between Ce^{3+} and Nd^{3+} ions has been previously proved by the superposition of the emission and absorption bands between the two ions [28–30,36–38]. Tai et al. demonstrated another possibility of energy transfer mechanism by near-infrared (NIR) quantum cutting involving the down-conversion of an absorbed visible photon to the emission of two NIR photons [36]. In Figure 2 the energy transfer mechanisms between Ce^{3+} and Nd^{3+} ions are depicted in an energy level diagram. When pump photons at wavelengths around 339 nm and 460 nm are absorbed, the Ce^{3+} ions are excited from the $^2F_{5/2}$ ground state to the broad pump bands $5d_2$ and $5d_1$, respectively. The electrons in the $5d_2$ ($^2B_{1g}$) pump band can relax non-radiatively to the lower $5d_1$ ($^2A_{1g}$) pump band and then decay radiatively to the $^2F_{5/2}$ ground state. The radiative transfer mechanism occurs between the transition $5d_1$ ($^2A_{1g}$) $\rightarrow$ $^2F_{5/2}$ of Ce^{3+} ion and the transition $^4I_{9/2} \rightarrow$ $^2G_{7/2}$ of Nd^{3+} ion, because of the strong overlap between the Ce^{3+} emission band and the Nd^{3+} absorption line, as shown in Figures 1 and 2, indicated by pathway (1) [28–30,36–38] through the cross-relaxation process. The other energy transfer mechanism based on the cooperative down-conversion is possible because the energy of the transition $5d_1$ ($^2A_{1g}$) $\rightarrow$ $^2F_{5/2}$ of Ce^{3+} ion is approximately twice as high as the energy difference between the $^4F_{3/2}$ and $^4I_{11/2}$ levels of the Nd^{3+} ion (Figure 2, as shown by pathway (2)) [36,39,40].

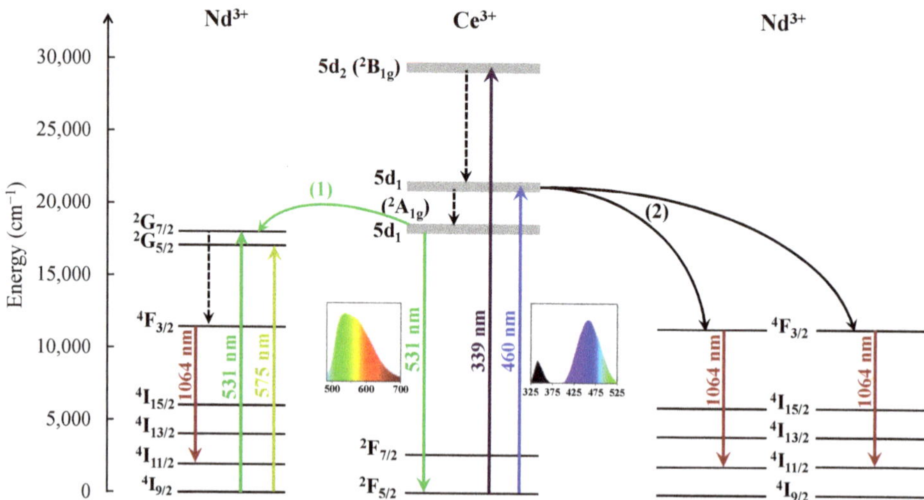

Figure 2. Energy-level diagram illustrating the energy transfer mechanisms between Ce^{3+} and Nd^{3+} ions in the Ce:Nd:YAG active medium (adapted from [36]). (1) Radiative energy transfer pathway. (2) Quantum cutting down-conversion pathway.

The good overlap between the Ce^{3+} absorption bands and the solar spectrum (Figure 1) in addition to the efficient energy transfer between Ce^{3+} and Nd^{3+} ions (Figure 2) make the Ce:Nd:YAG active medium of particular interest to solar-pumped laser researchers.

3. Solar-Pumped Ce:Nd:YAG Laser Systems

3.1. PROMES-CNRS Heliostat-Parabolic Mirror Solar Energy Collection and Concentration System

The medium size solar furnace (MSSF) of Procédés, Matériaux et Énergie Solaire—Centre National de la Recherche Scientifique (PROMES-CNRS) was used for the solar laser experiments. The incoming solar radiation was redirected by a large plane mirror (3.0 m × 3.0 m) with 36 small flat segments (0.5 m × 0.5 m each) mounted on a two-axis heliostat towards the stationary parabolic mirror with 2 m diameter, 60° rim angle, and 850 mm focal length (Figure 3). The external annular area of the parabolic mirror was masked to avoid overheating the gain medium so that only 1.38 m in diameter of its central circular area was used. An effective solar collection area of 1.09 m² was calculated after discounting the shading effects of a shutter, X-Y-Z axis positioner, solar laser cavity and 0.3 m diameter central opening of the parabolic mirror. The shutter was used to control the input power. All the mirrors of the MSSF solar facility are back-surface silver-coated, and because of the iron impurities within the glass substrates and imperfections, only 59% of incoming solar radiation was focused to the focal zone. Direct solar irradiances between 1036 W/m² and 1061 W/m² were measured during the experiments in Odeillo (France), in February 2022. A Kipp and Zonen CH1 pyrheliometer, on a Kipp and Zonen 2AP solar tracker, from Kipp and Zonen, Delft, The Netherlands, was used for the measurements.

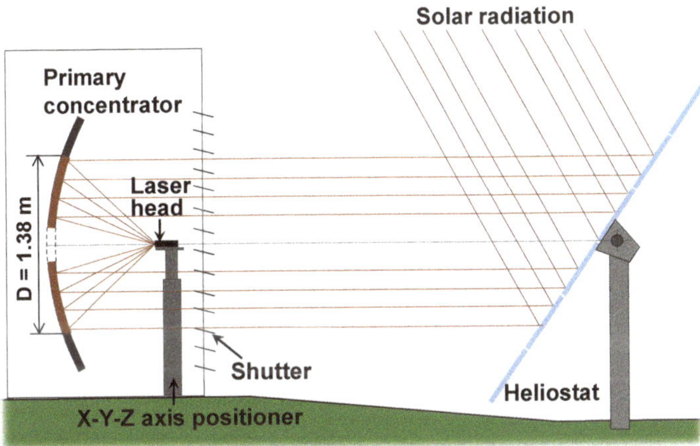

Figure 3. Schematics of the PROMES-CNRS solar energy collection and concentration system.

3.2. Side-Pumped Ce:Nd:YAG Solar Laser Heads

Two solar laser heads were investigated using as secondary concentrator: (1) a fused silica aspherical lens, and (2) a fused silica light guide with rectangular cross-section, as shown in Figures 4 and 5.

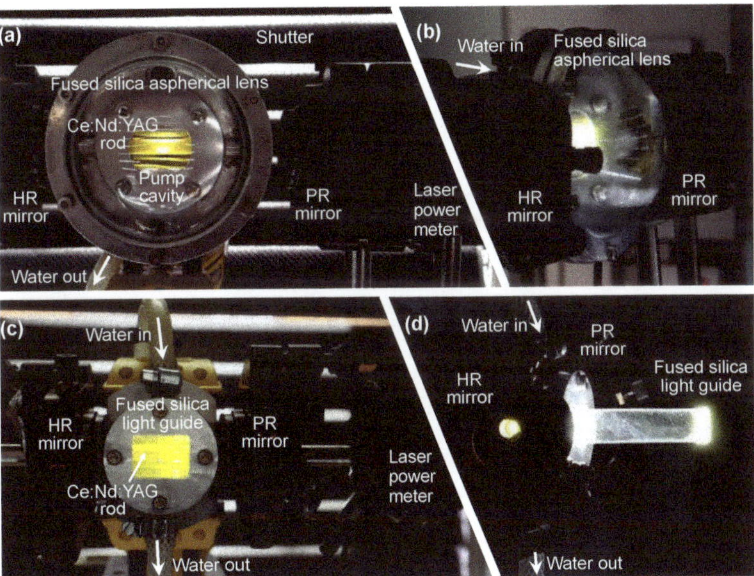

Figure 4. Photographs of the front view (**a**,**c**) and side view (**b**,**d**) of the Ce:Nd:YAG solar laser heads using the fused silica aspherical lens (**a**,**b**) and the fused silica light guide (**c**,**d**). PR: partial reflection; HR: high reflection.

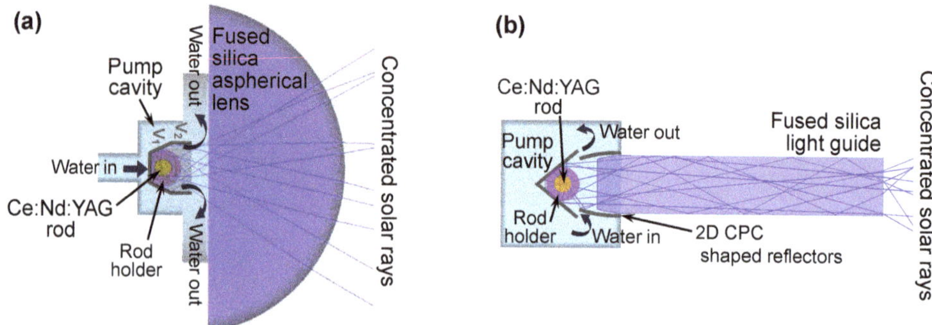

Figure 5. Schematics of the side-view of the solar laser heads with aspherical lens (**a**) and with light guide (**b**).

The aspherical lens of the first laser head was fabricated of fused silica material, which has a low coefficient of thermal expansion, being resistant to thermal shock, and a wide transparency range, which can extend from the UV into the NIR. It had 80 mm diameter, 60 mm radius of curvature, and 35 mm thickness. This laser head had also a tertiary concentrator constituted by a 2D pumping cavity, which had a trapezoidal-shaped reflector V_1 with a 25° opening angle, and two upper planar reflectors V_2 perpendicular to the bottom face of the pumping cavity (Figure 5a), enabling an effective coupling of the concentrated light rays with different incident angles into the rod [26]. This pumping cavity, covered with a 94%-reflectivity silver-coated aluminum foil, had an entrance aperture of 12.2 mm × 22.0 mm, a depth of 13.3 mm and an 11.0 mm separation between the plane output face of the aspherical lens and the laser rod optical axis.

The second laser head used a rectangular light guide also fabricated of fused silica material with 73.5 mm length, 14 mm × 18.4 mm input end, and an output section with the shape of a 2D CPC with 5.5 mm length and 12 mm × 22 mm output end. As tertiary concentrator, this laser head had a 2D V-shaped pump cavity with 16 mm × 22 mm entrance aperture and 10 mm depth, positioned at the exit of 2D CPC shaped reflectors with 12 mm × 22 mm output end (Figure 5b). All these provided an efficient coupling of the light to the laser rod. The inner walls of both the V-shaped and upper part of the pump cavity were bonded with a protected silver-coated aluminum foil with 94% reflectivity.

The active medium investigated was a Ce(0.1%):Nd(1.1%):YAG rod of 4.0 mm diameter and 35 mm length with both end faces polished and anti-reflection coated for the laser emission wavelength (reflectivity (R) < 0.2% @ 1064 nm), supplied by Chengdu Dongjun Laser Co., Ltd. (Chengdu, China). This rod had a refractive index = 1.8197 @ 1064 nm; perpendicularity < 3′, parallelism < 10″, flatness < 1/10λ; surface quality = 20/10; and damage threshold ≥ 5 J/cm^2 @ 1064 nm, 10 ns 10 Hz. The laser rod was mounted inside the pump cavity of each laser head. Both the cavity and the laser rod were actively cooled by water with 6 L/min flow rate.

Each solar laser head was fixed on the X-Y-Z axis positioner, ensuring a precise optical alignment in the focal zone. A high reflection (HR) coated flat mirror at 1064 nm (99.96% @ 1064 nm) and a partial reflection (PR) coated flat output coupler at 1064 nm (R ≥ 95% @ 1064 nm) constituted the optical resonator of short length. The laser output power was measured with a Thorlabs PM1100D power meter.

The laser beam quality M^2 factors were determined by measuring the beam diameter at $1/e^2$ at a near-field position (75 mm from the output coupler) and a far-field position (1.0 m away from the output coupler). The laser beam divergence θ was found using Equation (1):

$$\tan\theta = (\phi_2 - \phi_1) / 2L \tag{1}$$

where ϕ_1 and ϕ_2 are the measured laser beam diameters at $1/e^2$ width, 75 mm, and 1.0 m away from the output mirror, respectively, and L is the distance between these two points. The M^2 factor was then calculated by Equation (2):

$$M^2 = \theta/\theta_0 \qquad (2)$$

where $\theta_0 = \lambda/\pi\omega_0$ is the divergence of diffraction-limited Gaussian beam for $\lambda = 1.064$ μm and ω_0 is the beam waist radius, as calculated by LASCAD analysis for the 4 mm diameter rod within the symmetric laser resonator. This value can be confirmed by measuring the laser beam diameter at $1/e^2$ width ϕ_1 at near-field position.

4. Ce:Nd:YAG Side-Pumped Solar Laser Experiments with Either the Aspherical Lens or the Light Guide

Two side-pumped solar laser heads were investigated to study the influence of the secondary concentrators, aspherical lens and rectangular light guide, on the Ce:Nd:YAG output performance.

Aspherical lenses have a non-spherical surface, in which the radius of curvature gradually changes from the center of the lens to the edge, having a shape slightly divergent from spherical. The function of this asphere surface is to reduce or eliminate spherical aberration, an optical effect that causes incident light rays to focus on different points. Therefore, aspherical lenses can focus all the incident light on the exact same point, providing true diffraction-limited spot sizes and the lowest wavefront error. With aspherical lenses, the concentrated solar radiation can be collected and compressed efficiently from the focal zone of the parabolic mirror into the laser rod.

Rectangular light guides transmit the incident light through internal refractive and total internal reflection principles to a particular area. In solar-pumped lasers, they are used to distribute and uniform the concentrated solar radiation from the focal zone along the laser rod.

The main difference between these two secondary concentrators relies on the pump light distribution on the laser rod, as shown in Table 1. The absorbed pump light profile within the rod produced by the aspherical lens is more concentrated on its central region because this concentrator preserves the near-Gaussian profile of the pump source. The rectangular light guide enables a uniform absorbed pump light profile within the laser rod as it converts the near-Gaussian profile of the concentrated light source into a uniform rectangular light distribution on the output end. The uniformity of the light absorption on the rod helps to minimize the issues associated to the thermal load.

The Ce:Nd:YAG output performance from two solar laser heads was investigated in PROMES-CNRS, using the MSSF, during February 2022. For both, a short symmetric optical resonator with flat mirrors was used to extract the maximum laser power by the laser rod.

In Table 1 is summarized the output performance of both types of Ce:Nd:YAG side-pumped solar laser heads, and in Figure 6 is shown the laser power as a function of the incoming solar power. The solar laser with the aspherical lens produced 19.6 W laser power at 1064 nm with 1125 W of incoming solar power, corresponding to 18.0 W/m² collection efficiency. The solar laser with the light guide produced less laser power, 17.4 W; with slightly more incoming solar power, 1153 W; less solar-to-laser power conversion efficiency; and a higher threshold pump power than the one with the aspherical lens.

Table 1. Summary of the performance of the Ce:Nd:YAG side-pumped solar lasers with the aspherical lens and the rectangular light guide secondary concentrators.

	Aspherical Lens Secondary Concentrator	Light Guide Secondary Concentrator	Improvement of the Aspherical Lens over the Light Guide Concentrator (Times)
Scheme			
Absorbed pump profile			
Laser power (@1064 nm)	19.6 W	17.4 W	1.13
Threshold pump power	480 W	577 W	1.20
Collection efficiency	18.0 W/m^2	16.0 W/m^2	1.13
Slope efficiency	3.0%	3.0%	1.00
Solar-to-laser power conversion efficiency	1.7%	1.5%	1.15
M^2 factor (100% pump power)	30	38	—
M^2 factor (60% pump power)	20	35	—

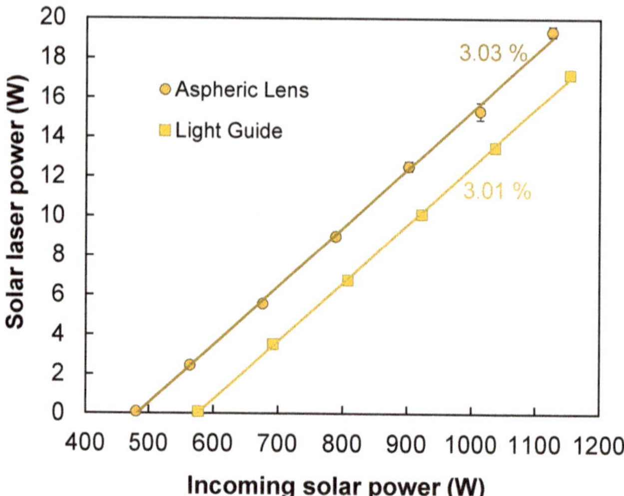

Figure 6. Solar laser power as a function of the incoming solar power of the Ce:Nd:YAG side-pumped solar lasers with the aspherical lens and the light guide secondary concentrators. Error bars, when not hidden by symbols, are for standard deviation.

The laser beam quality M^2 factors were calculated using Equations (1) and (2). $M_X^2 \approx M_Y^2$ factors of 38 and 35 were determined for the laser beam produced by the solar laser head with the light guide for 100% pump power and 60% of the total pump power, respectively. The M^2 factors measurement of the laser head with the aspherical lens was not possible, because the Ce:Nd:YAG rod fractured before laser beam quality measurements. Thus, they were calculated through numerical analysis by ZEMAX and LASCAD [20], resulting in $M_X^2 \approx M_Y^2$ factors of 30 and 20 for 100% pump power and 60% of the total pump power, respectively. The higher M^2 factors from the laser head with the light guide, compared with that with the aspherical lens was mainly due to the shorter resonator cavity length of 60 mm, in contrast to the 120 mm of cavity length used with the aspherical lens. The cavity lengths could not be equal in the two cases because of mechanical constraints in the laser head with the aspherical lens. Moreover, the beam quality M^2 factors deteriorated at higher pump power levels, as indicated in Table 1.

The greater laser efficiency produced by the solar laser head with the aspherical lens secondary concentrator can be mainly due to the higher concentration of the incident light in the central region of the laser rod provided by the aspherical lens. However, in an attempt to produce more laser power, by adding more input power, the Ce:Nd:YAG rod fractured on its central zone in the exact region where the solar light was mainly concentrated, as shown in Figure 7a. This event also occurred in a previous study where the Ce:Nd:YAG laser rod was used in an end-side-pumping configuration, in which the solar light was mostly focused on the top region of the rod [25]. This inhomogeneous pumping increased the thermal load of the rod in the region of more pumping intensity, which caused its final fracture when using higher solar powers, limiting the maximum laser output power.

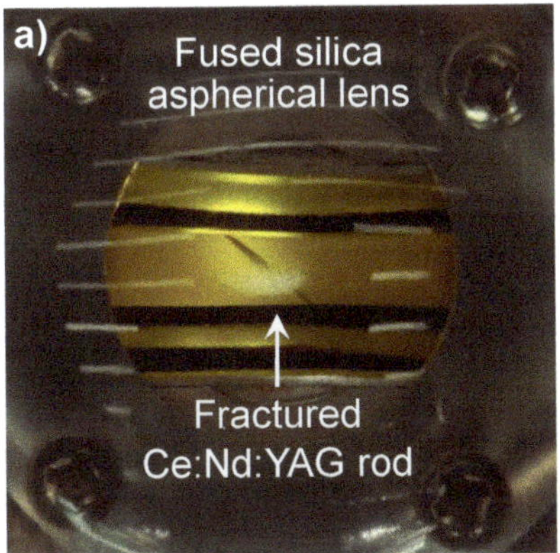

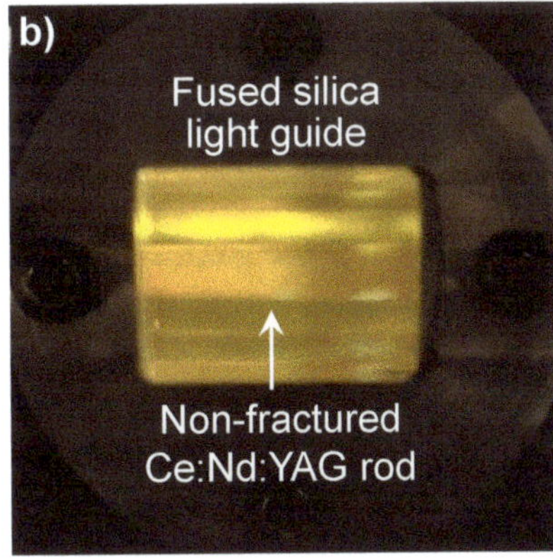

Figure 7. Photograph of the Ce:Nd:YAG laser rods on the laser heads with the fused liquid fused liquid aspheric lens (**a**) and rectangular light guide (**b**) after the solar laser experiments. The fracture on the Ce:Nd:YAG laser rod using the aspheric lens is visible on its central region.

The side-pumping configuration helps to minimize the thermal stress issues in the Ce:Nd:YAG active medium because it spreads the concentrated solar energy along the laser rod axis, which allows a greater increase in the output performance as already demonstrated [26]. Using an aspherical lens as a secondary concentrator enabled the increase in the output laser power; however, since the concentrated solar radiation was not evenly distributed along the laser rod, the absorbed pump distribution was still not uniform (Table 1), which led to Ce:Nd:YAG medium fracture (Figure 7a). The rectangular light guide was able to overcome this problem through the uniformity of the incident solar light on the rod. Although the laser output power was slightly inferior to that of the aspherical lens, the capacity to uniformly distribute the light led to a uniform absorbed pump distribution along the rod (Table 1). The thermal load on the rod was thus considerably reduced, which prevented the Ce:Nd:YAG medium fracture at higher input solar power levels (Figure 7b). Moreover, it also allowed a more stable laser emission during the two-week testing period in PROMES-CNRS without damaging the rod. Therefore, the fused silica light guide with rectangular cross-section is a promising secondary concentrator to enable the laser power scalability of the Ce:Nd:YAG solar laser.

The previous study, based on a Ce:Nd:YAG side-pumped solar laser with a semi-spherical lens as secondary concentrator, was accomplished by using the NOVA heliostat-parabolic mirror system with 75% reflectivity and 66 cm focal length [26], which is more efficient than the MSSF PROMES-CNRS facility with 59% reflectivity and 85 cm focal length.

Nevertheless, the Ce:Nd:YAG rod side-pumped through the NOVA solar facility produced only 16.5 W maximum continuous-wave power at 600 W incoming solar power [26]. In an attempt to boost the laser power by adding more solar power, the Ce:Nd:YAG rod was fractured. In the present study, by using the PROMES-CNRS facility, the Ce:Nd:YAG rod produced 19.6 W maximum continuous-wave laser power, which is 1.19 times more than that from the NOVA facility [26]. More importantly, the Ce:Nd:YAG rod was only fractured at 1125 W incoming solar power level, which is 1.88 times more than that of the previous side-pumped Ce:Nd:YAG laser rod could withstand at the NOVA solar facility. In conclusion, with the PROMES-CNRS solar facility, the solar laser head with aspherical lens offered a much more thermal resistance and a higher solar laser power compared with that obtained with the NOVA facility [26].

It is important to note that the Ce:Nd:YAG active medium is more susceptible to thermal fracture than the Nd:YAG medium when pumped by sunlight. The main reason relies on the Stokes efficiency, which is lower for the Ce:Nd:YAG rod, 0.43 [27] compared with the 0.62 of the Nd:YAG rod [7]. Such low Stokes efficiency results in loss of pump photon energy normally in the form of heat, increasing the thermal load in the rod. Therefore, it is important to design solar laser systems that take this factor into account. The pumping distribution on the active medium plays an important role in the optimization process of scaling solar-pumped lasers to high power.

5. Conclusions

Two solar laser heads with two different secondary concentrators were investigated to study the influence of their pumping distribution on the Ce:Nd:YAG side-pumped solar laser output performance. The solar laser head with the fused silica aspherical lens produced a non-uniform absorbed pump profile within the rod, which is more concentrated on its central region, while the solar laser head with the rectangular fused silica light guide formed a uniform absorbed pump profile along the rod. The laser collection efficiency, solar-to-laser power conversion efficiency, and threshold pump power with the aspherical lens were 18 W/m^2, 1.7%, and 480 W, respectively, which were better than that with the light guide that had 16 W/m^2, 1.5%, and 577 W, respectively. However, the non-uniformity of the absorbed pump profile created by the aspherical lens increased the thermal load of the rod, creating hot pump spots. This led to the Ce:Nd:YAG rod fracture, limiting the maximum laser output power. At the expense of slightly lower solar laser power and efficiency, the use of light guide overcame this problem through the uniformity of the incident solar light on the rod. Therefore, the pumping distribution plays an important role in the design of Ce:Nd:YAG solar laser systems, and the light guide technology may ensure successful solar laser power scaling.

Author Contributions: Conceptualization, C.R.V. and D.L; methodology, C.R.V.; validation, C.R.V. and M.C.; formal analysis, C.R.V. and D.G.; investigation, C.R.V., D.L., J.A. and E.G.; resources, D.L., J.A. and E.G.; data curation, C.R.V., H.C. and B.D.T.; writing—original draft preparation, C.R.V.; writing—review and editing, C.R.V., D.L., J.A., D.G, M.C., B.D.T. and H.C.; supervision, D.L.; project administration, D.L.; funding acquisition, D.L., J.A. All authors have read and agreed to the published version of the manuscript.

Funding: This research was funded by Science and Technology Foundation of Portuguese Ministry of Science, Technology and Higher Education (FCT—MCTES), through the strategic project UIDB/00068/2020 and the exploratory research project EXPL/FIS-OTI/0332/2021. This research was also funded by the Solar Facilities for European Research Area—Third Phase (SFERA III), Grant Agreement No. 823802.

Institutional Review Board Statement: Not applicable.

Informed Consent Statement: Not applicable.

Data Availability Statement: Not applicable.

Acknowledgments: The fellowship grants PD/BD/128267/2016, PD/BD/142827/2018, SFRH/BD/145322/2019, SFRH/BPD/125116/2016, 2021.06172.BD, and the contract CEECIND/03081/2017 of Cláudia R. Vistas, Dário Garcia, Miguel Catela, Bruno D. Tibúrcio, Hugo Costa and Joana Almeida are acknowledged.

Conflicts of Interest: The authors declare no conflict of interest.

References

1. Lando, M.; Kagan, J.A.; Shimony, Y.; Kalisky, Y.Y.; Noter, Y.; Yogev, A.; Rotman, S.R.; Rosenwaks, S. *Solar-Pumped Solid State Laser Program*; SPIE: Bellingham, WA, USA, 1997; Volume 3110.
2. Vasile, M.; Maddock, C.A. Design of a formation of solar pumped lasers for asteroid deflection. *Adv. Space Res.* **2012**, *50*, 891–905. [CrossRef]
3. Yamada, N.; Ito, T.; Ito, H.; Motohiro, T.; Takeda, Y. Crystalline silicon photovoltaic cells used for power transmission from solar-pumped lasers: III. Prototype for proof of the light trapping concept. *Jpn. J. Appl. Phys.* **2018**, *57*, 08RF07. [CrossRef]
4. Kiss, Z.J.; Lewis, H.R.; Duncan, R.C. Sun pumped continuous optical maser. *Appl. Phys. Lett.* **1963**, *2*, 93–94. [CrossRef]
5. Young, C.G. A Sun-Pumped cw One-Watt Laser. *Appl. Opt.* **1966**, *5*, 993–997. [CrossRef] [PubMed]
6. Arashi, H.; Oka, Y.; Sasahara, N.; Kaimai, A.; Ishigame, M. A Solar-Pumped cw 18 W Nd:YAG Laser. *Jpn. J. Appl. Phys.* **1984**, *23*, 1051–1053. [CrossRef]
7. Weksler, M.; Shwartz, J. Solar-pumped solid-state lasers. *IEEE J. Quantum Electron.* **1988**, *24*, 1222–1228. [CrossRef]
8. Lando, M.; Kagan, J.; Linyekin, B.; Dobrusin, V. A solar-pumped Nd:YAG laser in the high collection efficiency regime. *Opt. Commun.* **2003**, *222*, 371–381. [CrossRef]
9. Dinh, T.H.; Ohkubo, T.; Yabe, T.; Kuboyama, H. 120 watt continuous wave solar-pumped laser with a liquid light-guide lens and an Nd:YAG rod. *Opt. Lett.* **2012**, *37*, 2670–2672. [CrossRef]
10. Liang, D.; Vistas, C.R.; Tiburcio, B.D.; Almeida, J. Solar-pumped Cr:Nd:YAG ceramic laser with 6.7% slope efficiency. *Sol. Energy Mater. Sol. Cells* **2018**, *185*, 75–79. [CrossRef]
11. Ohkubo, T.; Yabe, T.; Yoshida, K.; Uchida, S.; Funatsu, T.; Bagheri, B.; Oishi, T.; Daito, K.; Ishioka, M.; Nakayama, Y.; et al. Solar-pumped 80 W laser irradiated by a Fresnel lens. *Opt. Lett.* **2009**, *34*, 175–177. [CrossRef]
12. Almeida, J.; Liang, D.; Vistas, C.R.; Guillot, E. Highly efficient end-side-pumped Nd:YAG solar laser by a heliostat-parabolic mirror system. *Appl. Opt.* **2015**, *54*, 1970–1977. [CrossRef] [PubMed]
13. Liang, D.; Almeida, J.; Vistas, C.R. 25 W/m^2 collection efficiency solar-pumped Nd:YAG laser by a heliostat-parabolic mirror system. *Appl. Opt.* **2016**, *55*, 7712–7717. [CrossRef] [PubMed]
14. Liang, D.; Almeida, J.; Vistas, C.R.; Guillot, E. Solar-pumped Nd:YAG laser with 31.5 W/m^2 multimode and 7.9 W/m^2 TEM00-mode collection efficiencies. *Sol. Energy Mater. Sol. Cells* **2017**, *159*, 435–439. [CrossRef]
15. Guan, Z.; Zhao, C.; Zhang, H.; Li, J.; He, D.; Almeida, J.; Vistas, C.R.; Liang, D. 5.04% system slope efficiency solar-pumped Nd:YAG laser by a heliostat-parabolic mirror system. *J. Photonics Energy* **2018**, *8*, 027501. [CrossRef]
16. Guan, Z.; Zhao, C.; Li, J.; He, D.; Zhang, H. 32.1 W/m^2 continuous wave solar-pumped laser with a bonding Nd:YAG/YAG rod and a Fresnel lens. *Opt. Laser Technol.* **2018**, *107*, 158–161. [CrossRef]
17. Vistas, C.R.; Liang, D.; Almeida, J.; Tibúrcio, B.D.; Garcia, D. A doughnut-shaped Nd:YAG solar laser beam with 4.5 W/m^2 collection efficiency. *Sol. Energy* **2019**, *182*, 42–47. [CrossRef]
18. Liang, D.; Almeida, J. Solar-Pumped TEM$_{00}$ Mode Nd:YAG laser. *Opt. Express* **2013**, *21*, 25107–25112. [CrossRef]
19. Liang, D.; Almeida, J.; Vistas, C.R.; Guillot, E. Solar-pumped TEM$_{00}$ mode Nd:YAG laser by a heliostat-Parabolic mirror system. *Sol. Energy Mater. Sol. Cells* **2015**, *134*, 305–308. [CrossRef]
20. Vistas, C.R.; Liang, D.; Almeida, J.; Guillot, E. TEM$_{00}$ mode Nd:YAG solar laser by side-pumping a grooved rod. *Opt. Commun.* **2016**, *366*, 50–56. [CrossRef]
21. Bouadjemine, R.; Liang, D.; Almeida, J.; Mehellou, S.; Vistas, C.R.; Kellou, A.; Guillot, E. Stable TEM$_{00}$-mode Nd:YAG solar laser operation by a twisted fused silica light-guide. *Opt. Laser Technol.* **2017**, *97*, 1–11. [CrossRef]
22. Mehellou, S.; Liang, D.; Almeida, J.; Bouadjemine, R.; Vistas, C.R.; Guillot, E.; Rehouma, F. Stable solar-pumped TEM$_{00}$-mode 1064 nm laser emission by a monolithic fused silica twisted light guide. *Sol. Energy* **2017**, *155*, 1059–1071. [CrossRef]
23. Liang, D.; Vistas, C.R.; Almeida, J.; Tiburcio, B.D.; Garcia, D. Side-pumped continuous-wave Nd:YAG solar laser with 5.4% slope efficiency. *Sol. Energy Mater. Sol. Cells* **2019**, *192*, 147–153. [CrossRef]
24. Yagi, H.; Yanagitani, T.; Yoshida, H.; Nakatsuka, M.; Ueda, K. The optical properties and laser characteristics of Cr^{3+} and Nd^{3+} co-doped Y3Al5O12 ceramics. *Opt. Laser Technol.* **2007**, *39*, 1295–1300. [CrossRef]
25. Vistas, C.R.; Liang, D.; Garcia, D.; Almeida, J.; Tibúrcio, B.D.; Guillot, E. Ce:Nd:YAG continuous-wave solar-pumped laser. *Optik* **2020**, *207*, 163795. [CrossRef]
26. Vistas, C.R.; Dawei, L.; Joana, A.; Bruno, D.T.; Dário, G.; Miguel, C.; Hugo, C.; Emmanuel, G. Ce:Nd:YAG side-pumped solar laser. *J. Photonics Energy* **2021**, *11*, 018001. [CrossRef]
27. Villars, B.; Steven Hill, E.; Durfee, C.G. Design and development of a high-power LED-pumped Ce:Nd:YAG laser. *Opt. Lett.* **2015**, *40*, 3049–3052. [CrossRef]

28. Meng, J.X.; Li, J.Q.; Shi, Z.P.; Cheah, K.W. Efficient energy transfer for Ce to Nd in Nd/Ce codoped yttrium aluminum garnet. *Appl. Phys. Lett.* **2008**, *93*, 221908. [CrossRef]

29. Samuel, P.; Yanagitani, T.; Yagi, H.; Nakao, H.; Ueda, K.I.; Babu, S.M. Efficient energy transfer between Ce^{3+} and Nd^{3+} in cerium codoped Nd: YAG laser quality transparent ceramics. *J. Alloys Compd.* **2010**, *507*, 475–478. [CrossRef]

30. Li, Y.; Zhou, S.; Lin, H.; Hou, X.; Li, W. Intense 1064nm emission by the efficient energy transfer from Ce^{3+} to Nd^{3+} in Ce/Nd co-doped YAG transparent ceramics. *Opt. Mater.* **2010**, *32*, 1223–1226. [CrossRef]

31. Guo, Y.; Huang, J.; Ke, G.; Ma, Y.; Quan, J.; Yi, G. Growth and optical properties of the Nd, Ce:YAG laser crystal. *J. Lumin.* **2021**, *236*, 118134. [CrossRef]

32. Payziyev, S.; Sherniyozov, A.; Bakhramov, S.; Zikrillayev, K.; Khalikov, G.; Makhmudov, K.; Ismailov, M.; Payziyeva, D. Luminescence sensitization properties of Ce: Nd: YAG materials for solar pumped lasers. *Opt. Commun.* **2021**, *499*, 127283. [CrossRef]

33. *ASTM G173-03*; Standard tables for reference solar spectral irradiances: Direct normal and hemispherical on 37° tilted surface. ASTM International: West Conshohocken, PA, USA, 2012.

34. Nishiura, S.; Tanabe, S.; Fujioka, K.; Fujimoto, Y. Properties of transparent Ce:YAG ceramic phosphors for white LED. *Opt. Mater.* **2011**, *33*, 688–691. [CrossRef]

35. Prahl, S. Nd:YAG—Nd:Y3Al5O12. Available online: https://omlc.org/spectra/lasermedia/html/052.html (accessed on 5 April 2022).

36. Tai, Y.; Zheng, G.; Wang, H.; Bai, J. Near-infrared quantum cutting of Ce^{3+}–Nd^{3+} co-doped $Y_3Al_5O_{12}$ crystal for crystalline silicon solar cells. *J. Photochem. Photobiol. A* **2015**, *303*, 80–85. [CrossRef]

37. Mares, J.; Jacquier, B.; Pédrini, C.; Boulon, G. Energy transfer mechanisms between Ce^{3+} and Nd^{3+} in YAG: Nd, Ce at low temperature. *Rev. Phys. Appl.* **1987**, *22*, 145–152. [CrossRef]

38. Yamaga, M.; Oda, Y.; Uno, H.; Hasegawa, K.; Ito, H.; Mizuno, S. Energy transfer from Ce to Nd in $Y_3Al_5O_{12}$ ceramics. *Phys. Status Solidi C* **2012**, *9*, 2300–2303. [CrossRef]

39. Wegh René, T.; Donker, H.; Oskam Koenraad, D.; Meijerink, A. Visible Quantum Cutting in $LiGdF_4:Eu^{3+}$ Through Downconversion. *Science* **1999**, *283*, 663–666. [CrossRef]

40. Liu, X.; Teng, Y.; Zhuang, Y.; Xie, J.; Qiao, Y.; Dong, G.; Chen, D.; Qiu, J. Broadband conversion of visible light to near-infrared emission by Ce^{3+}, Yb^{3+}-codoped yttrium aluminum garnet. *Opt. Lett.* **2009**, *34*, 3565–3567. [CrossRef]

Article

40 W Continuous Wave Ce:Nd:YAG Solar Laser through a Fused Silica Light Guide

Joana Almeida [1], Dawei Liang [1,*], Dário Garcia [1], Bruno D. Tibúrcio [1], Hugo Costa [1], Miguel Catela [1], Emmanuel Guillot [2] and Cláudia R. Vistas [1]

[1] Centro de Física e Investigação Tecnológica (CEFITEC), Departamento de Física, Faculdade de Ciências e Tecnologia, Universidade NOVA de Lisboa, 2829-516 Caparica, Portugal; jla@fct.unl.pt (J.A.); dm.garcia@campus.fct.unl.pt (D.G.); brunotiburcio78@gmail.com (B.D.T.); hf.costa@campus.fct.unl.pt (H.C.); m.catela@campus.fct.unl.pt (M.C.); c.vistas@fct.unl.pt (C.R.V.)

[2] PROMES-CNRS, 7 rue du Four Solaire, 66120 Font-Romeu-Odeillo-Via, France; emmanuel.guillot@promes.cnrs.fr

* Correspondence: dl@fct.unl.pt

Abstract: The solar laser power scaling potential of a side-pumped Ce:Nd:YAG solar laser through a rectangular fused silica light guide was investigated by using a 2 m diameter parabolic concentrator. The laser head was formed by the light guide and a V-shaped pump cavity to efficiently couple and redistribute the concentrated solar radiation from the parabolic mirror to a 4 mm diameter, 35 mm length Ce(0.1 at.%):Nd(1.1 at.%):YAG laser rod. The rectangular light guide ensured a homogeneous distribution of the solar radiation along the laser rod, allowing it to withstand highly concentrated solar energy. With the full collection area of the parabolic mirror, the maximum continuous wave (cw) solar laser power of 40 W was measured. This, to the best of our knowledge, corresponds to the highest cw laser power obtained from a Ce:Nd:YAG medium pumped by solar radiation, representing an enhancement of two times over that of the previous side-pumped Ce:Nd:YAG solar laser and 1.19 times over the highest Cr:Nd:YAG solar laser power with a rectangular light-guide. This research proved that, with an appropriate pumping configuration, the Ce:Nd:YAG medium is very promising for scaling solar laser output power to a higher level.

Keywords: Ce:Nd:YAG; solar laser; light-guide; homogenizer; side-pumped; parabolic mirror

Citation: Almeida, J.; Liang, D.; Garcia, D.; Tibúrcio, B.D.; Costa, H.; Catela, M.; Guillot, E.; Vistas, C.R. 40 W Continuous Wave Ce:Nd:YAG Solar Laser through a Fused Silica Light Guide. *Energies* **2022**, *15*, 3998. https://doi.org/10.3390/en15113998

Academic Editor: Albert Ratner

Received: 12 May 2022
Accepted: 27 May 2022
Published: 29 May 2022

Publisher's Note: MDPI stays neutral with regard to jurisdictional claims in published maps and institutional affiliations.

1. Introduction

Solar-powered laser systems directly convert broadband and incoherent solar radiation into narrowband and coherent laser radiation through an active medium. Since this technology can be operated using only renewable energy, it may bring an important economic advantage for countries with high solar availability [1] and for the future development of sustainable industrialization [2], either on Earth [3,4] or in Space [5,6].

The first laser emission achieved by pumping an active medium with solar energy was reported by Kiss et al. in 1963, based on a calcium fluoride crystal doped with a divalent dysprosium (Dy^{2+}:CaF_2) medium, reaching the continuous wave (cw) laser action at 2.36 µm [7]. Thenceforth, optical pumping designs and active media have been investigated for solar-pumped lasers. Between the late 1970s and early 2000s, gas [6,8,9], liquid [10], and solid [11–13] active media have all been evaluated as potential candidates for solar-pumped lasers. Still, this research has essentially converged in the use of bulk solid-state optical gain media, namely the yttrium aluminum garnet ($Y_3Al_5O_{12}$) doped with the rare earth ion neodymium (Nd^{3+}) [14–27]. The favorable spectroscopic characteristics of the Nd^{3+} active ion and optomechanical properties of the YAG host material [28], in conjunction with the advances in the optical pumping designs, have contributed to the progress of solar-pumped lasers performances [14–24]. Nevertheless, Nd^{3+} is not an ideal dopant for solar-pumped lasers due to the low spectral overlap of the Nd^{3+} absorption spectrum with the blackbody-like solar emission spectrum, imposing limits to the efficiency of solar-powered lasers.

Co-doping the Nd:YAG medium with chromium (Cr^{3+}) or cerium (Ce^{3+}) ions may improve the solar laser efficiency by supplying a broader absorption band to overlap with the solar emission spectrum, when compared to simple Nd:YAG lasers [29–32]. The attempts to meliorate the solar laser efficiency using Cr:Nd:YAG have been carried out since 2007 [22–24]. The record in solar laser slope efficiency, obtained by measuring the laser output power variation with the incoming solar power, is 6.7%, reached in 2018 by end-side-pumping a Cr(0.1 at.%):Nd(1.0 at.%):YAG rod through a primary parabolic concentrator [24]. This result is 1.28 times more than the highest slope efficiency obtained by Nd:YAG laser rod end-side-pumped through a similar solar facility [16].

The mechanisms of energy transfer between Ce^{3+} and Nd^{3+} ions have also been studied for some time using dye lasers [33], lamps [34–36], LEDs [37], and laser diodes [38] as pumping sources. Still, the experimental evaluation of the Ce:Nd:YAG performance under broadband solar pumping is in its early stages [25–27]. The first solar laser experiment using a Ce:Nd:YAG laser medium was performed in 2020, by end-side-pumping a Ce(0.05 at.%):Nd(1.0 at.%):YAG rod of 5 mm in diameter and 30 mm in length at the medium size solar furnace (MSSF) of PROMES-CNRS [25]. The Ce:Nd:YAG rod emitted a 6.0 W cw laser power at an incoming solar power of 964 W. However, at higher solar power levels, the Ce:Nd:YAG laser power dropped abruptly, followed by a fracture of the laser rod in its upper-end region, where most of the solar rays were focused [25]. To try to minimize this problem, side-pumping configurations were adopted in the most recent Ce:Nd:YAG solar lasers [26,27]. Side-pumping is suitable for laser power scaling since it spreads the concentrated solar energy along the laser rod axis, minimizing the thermal load problems. For 1125 W of incoming solar power at the MSSF of PROMES-CNRS, the maximum cw solar laser power of 19.6 W was emitted by a side-pumped 4 mm diameter and 35 mm length Ce(0.1 at.%):Nd(1.1 at.%):YAG laser rod [27], corresponding to an increase of 3.27 times in relation to that of the previous end-side-pumped Ce:Nd:YAG solar laser through the same solar facility [25]. For efficient side-pumping of the Ce:Nd:YAG laser rod, a fused silica aspherical lens was used as a secondary concentrator [27]. However, this resulted in an uneven light distribution along the laser rod, which led to its fracture when adding more incoming solar power.

To scale the Ce:Nd:YAG solar laser to higher power levels, a uniformly pumped laser rod is extremely important. The use of side-pumping configurations employing fused silica light guides with a rectangular cross-section has allowed a substantial reduction of the thermal load and stress problems in either Nd:YAG or Cr:Nd:YAG laser rods [20–22]. Based on the principles of total internal reflection, the rectangular light guides shaped the near-Gaussian distribution of the concentrated sunlight spot incident on its input aperture into a homogeneous light distribution at its output aperture [20–22]. Consequently, the solar radiation was uniformly spread along the laser rod, significantly reducing the accumulated heat within the laser medium compared to previous solar lasers with no light guide [16,24–27]. The highest laser output power attained by side-pumping a laser rod with a light guide was 33.6 W, from a Cr:Nd:YAG laser rod of 7 mm in diameter and 30 mm in length, pumped through the 2 m diameter MSSF parabolic mirror [22].

Due to the abovementioned reasons, we decided to test the laser power scalability of a 4 mm diameter, 35 mm length Ce(0.1 at.%):Nd(1.1 at.%):YAG laser rod side-pumped by a rectangular fused silica light guide. The experiments were realized at the MSSF of PROMES-CNRS. The side-pumping configuration with the fused silica light guide allowed the Ce:Nd:YAG laser rod to withstand an elevated incoming solar power level of 2.6 kW, producing a 40 W cw solar laser power with the full collection area of the MSSF parabolic mirror. This, as far as we are aware, was the highest reported cw solar laser power emitted from a Ce:Nd:YAG solar laser medium. It was also the highest cw solar laser power achieved with side-pumping configuration using a light homogenizer.

2. Energy Transfer Mechanism of Ce^{3+} to Nd^{3+} in Ce:Nd:YAG Medium

Although reports on solar laser using Ce:Nd:YAG as a gain medium have appeared only recently [25–27], the effect of co-doping the Nd:YAG medium with Ce^{3+} ions has stimulated already decades of research [33–38]. The strong and broad absorption bands of Ce:Nd:YAG make this active material a suitable candidate for broadband pumped lasers. As observed in Figure 1, the absorption spectrum of Ce:Nd:YAG presents two broad bands characteristic of the Ce^{3+} ion in a YAG lattice, which are centered at 339 nm and 460 nm (Figure 1b) and overlap well with the solar spectrum region of greater intensity (Figure 1a). It has also other bands in the red and near infrared (NIR) regions that are characteristic of the Nd^{3+} ions (Figure 1c).

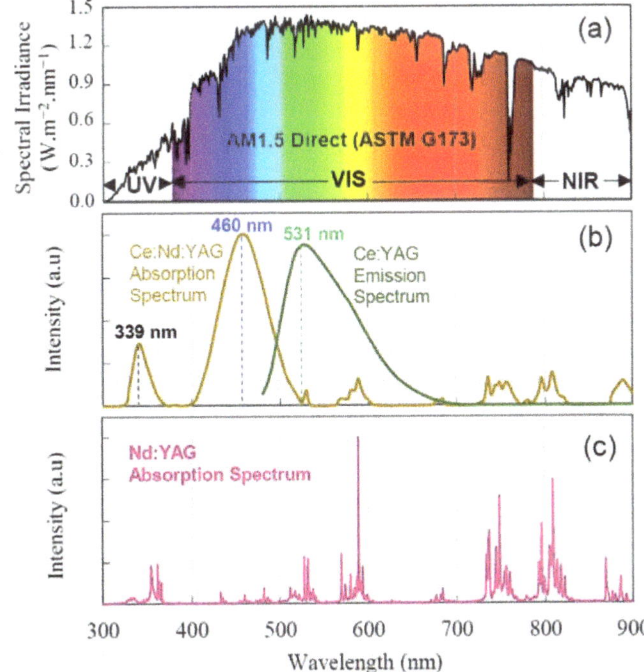

Figure 1. (**a**) AM1.5 direct solar spectrum at 300–900 nm wavelength range, adapted from [39]; UV, ultraviolet; VIS, visible; NIR, near infrared; (**b**) Ce:Nd:YAG absorption spectrum and Ce:YAG fluorescence spectrum, adapted from [35,40], respectively; (**c**) Nd:YAG absorption spectrum, adapted from [41].

Figure 2 illustrates the energy level diagram of Ce:Nd:YAG and the transfer mechanism between the Ce^{3+} and Nd^{3+} ions. When pumping Ce:Nd:YAG with broadband radiation, with wavelengths centered around 339 nm and 460 nm, the Ce^{3+} ions are excited from the ^{2}F$_{5/2}$ ground state level to the pump bands of 5d$_2$ (^{2}B$_{1g}$) and 5d$_1$ (^{2}A$_{1g}$), respectively. The excited electrons in the 5d$_2$ (^{2}B$_{1g}$) pump band then relax non-radiatively (dashed black line) to the lower 5d$_1$ pump band and further decay radiatively to the ^{2}F$_{5/2}$ ground state (solid green line). Since the absorption spectrum of Ce:Nd:YAG is quite broad, a strong and broad emission luminescence centered at 531 nm occurs (Figure 1b), which overlaps with the strong peak absorptions of Nd^{3+} in the green and yellow spectral region (Figure 1c). Therefore, a radiative transfer mechanism takes place between the 5d$_1$(^{2}A$_{1g}$) $\rightarrow$ ^{2}F$_{5/2}$ and ^{4}I$_{9/2}$ $\rightarrow$ ^{2}G$_{7/2}$ transitions of Ce^{3+} and Nd^{3+} ions, respectively, through a cross-relaxation process [33–35,38], as indicated by the pathway (1) of Figure 2.

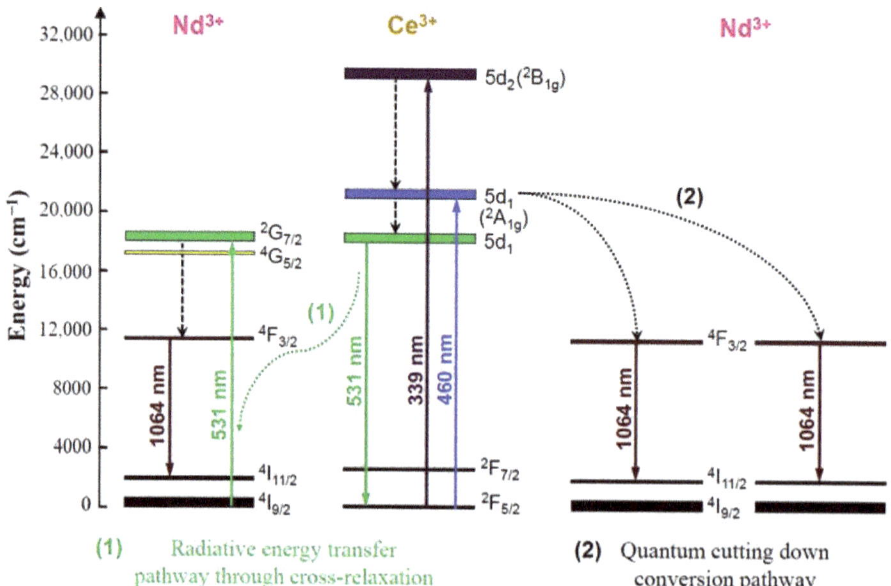

Figure 2. Energy-level diagram representing the energy transfer between Ce^{3+} and Nd^{3+} ions in the Ce:Nd:YAG material, adapted from [35]. (1) Radiative energy transfer pathway through cross-relaxation; (2) Quantum cutting down conversion pathway.

Another energy transfer mechanism may also occur, based on a quantum cutting down-conversion process [35,42], as represented by pathway (2). This process implies that two NIR photons could emit from one absorbed visible (VIS) photon. Its occurrence is possible in Ce:Nd:YAG due to the excitation of Ce^{3+} to the higher $5d_1(^2A_{1g})$ energy level, whose transition to the $^2F_{5/2}$ ground-state level has approximately twice the energy difference between the $^4F_{3/2}$ and $^4I_{11/2}$ levels of the Nd^{3+}.

3. Description of the Ce:Nd:YAG Solar Laser System

3.1. Medium Size Solar Furnace of the PROMES-CNRS

The solar facility is constituted by a heliostat and a stationary parabolic concentrator with horizontal axis. The heliostat has 36 flat mirror segments, each with 0.5 m × 0.5 m dimensions, which tracks and redirects the solar rays to the parabolic mirror with a 2.0 m diameter, a 0.85 m focal distance, and a 60° rim angle [16,20–22,25], as illustrated in Figure 3a. Both heliostat and parabolic mirrors are back-silvered, each with less than 80% reflectivity due to iron contents in the glass substrate and degradation owing to many years of usage. Hence, about 59% of the incoming solar rays are focused [16,21,22,25], reaching a maximum of 2 kW of power at the focus [43].

For the solar laser experiments, the Ce:Nd:YAG solar laser head, along with the resonant cavity, was fixed on a mechanical support with automatic X-Y-Z axis calibration at the focal zone of the MSSF, as shown in Figure 3b. The incoming solar power was regulated by a shutter with motorized blades. When the shutter was totally open, a 2.48 m² effective collection area was measured for the maximum diameter of the parabolic mirror. All the shadow effects in the MSSF caused by the space between the flat mirrors of the heliostat, the shutter blades, the laser head, and respective mechanical supports, have been accounted for.

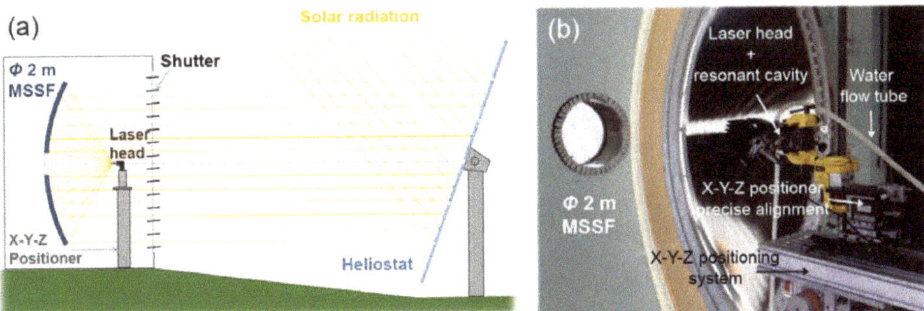

Figure 3. (**a**) Design of the PROMES-CNRS solar facility. (**b**) Photograph of the Medium Size Solar Furnace (MSSF) and the solar laser head mounting.

3.2. Side-Pumped Solar Laser Head with the Fused Silica Homogenizer

A schematic design of the laser head is given in Figure 4. It is composed of a rectangular fused silica light guide and a two-dimensional (2D) V-shaped pump cavity, within which the Ce:Nd:YAG laser rod was fixed, being actively cooled by water.

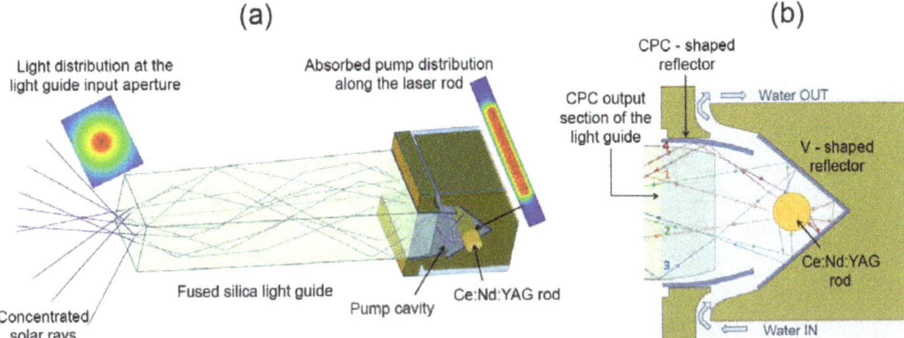

Figure 4. Schematic designs of (**a**) the Ce:Nd:YAG solar laser head with rectangular light guide and (**b**) the V-shaped pump cavity with the passage of pump rays within the laser rod. The insets of (**a**) represent the solar rays' distribution at both the light guide input aperture (**left**) and along the laser rod (**right**).

Fused silica material is suitable for solar-powered lasers since it has a wide transparency range over the Ce:Nd:YAG absorption spectrum. Furthermore, it is very resistant to high temperature and thermal shock [44]. The light guide with a 14 mm × 18.4 mm input aperture collected the MSSF-concentrated solar radiation with a near-Gaussian profile of 11 mm full width at half maximum [16,21,25]. The solar rays were then internally reflected along a 68 mm length of the light guide, to its 2D compound parabolic concentrator (CPC) output section with 12 mm × 22 mm output aperture and 5.5 mm length. As demonstrated in Figure 4a, the fused silica light guide behaves like a beam homogenizer, enabling a homogeneous absorbed pump light distribution along the Ce:Nd:YAG laser rod. This helps not only to reduce the thermal effects in the laser rod, but also to compensate the heliostat tracking error-dependent losses [20–22], leading to higher solar laser beam stability compared to solar lasers with no light homogenizer [45].

To produce the light guide, a fused silica slab of 99.995% optical purity with 14 mm × 22 mm × 75 mm dimensions (supplied by Beijing Aomolin Ltd., Beijing, China), was ground and polished to the final dimensions abovementioned. The side surfaces of the light guide were slightly inclined to ensure its easy mechanical attachment to the laser head. The 2D-CPC sidewalls of the light guide provided effective coupling of the solar rays to the laser rod. It is also worth noting that the direct cooling of the laser rod with water is of utmost importance to prevent the laser rod from UV and IR heating. A water flow rate of about 6 L/min was adopted in the present work, which was essential to dissipate the heat within the laser head.

The water-flooded V-shaped pump cavity had a 10 mm depth and a 16 mm × 22 mm entrance aperture, positioned at the exit of the 2D CPC-shaped upper reflectors with a 12 mm × 22 mm output aperture. This arrangement provided the zigzag path of the solar rays within the Ce:Nd:YAG laser rod. As demonstrated in Figure 4b, ray one (orange color) hits directly the laser medium and is redirected back to the rod by the V-shaped cavity so that double-pass pumping is accomplished. The solar rays from the light guide that do not directly reach the laser rod, represented by rays two (green color) and three (blue color), can be redirected again to the laser rod by the V-shaped reflector. The upper 2D CPC-shaped reflectors also help to redirect the rays that exit the 2D CPC surface of the light-guide to the laser rod, as demonstrated by the optical path of ray four (red color), which may pass through the rod twice with the help of the V-shaped cavity. The inner walls of the V-shaped section, as well as the upper part section of the pump cavity, were covered with silver-coated aluminum foils of 94% reflectivity.

A high reflection (HR) coated mirror (99.9% @ 1064 nm), the Ce:Nd:YAG laser rod, and a partial reflection (PR) coated mirror (R ≥ 95 ± 2% @ 1064 nm) formed the optical resonator, as shown in Figure 5. The 4 mm diameter and 35 mm length Ce(0.1 at.%):Nd(1.1 at.%):YAG rod was manufactured by Chengdu Dongjun Laser Co., Ltd. The end faces of the laser rod were covered with anti-reflective coating for 1064 nm (reflectivity (R) < 0.2% @ 1064 nm). Both HR and PR laser mirrors were supplied by ESKMA Optics.

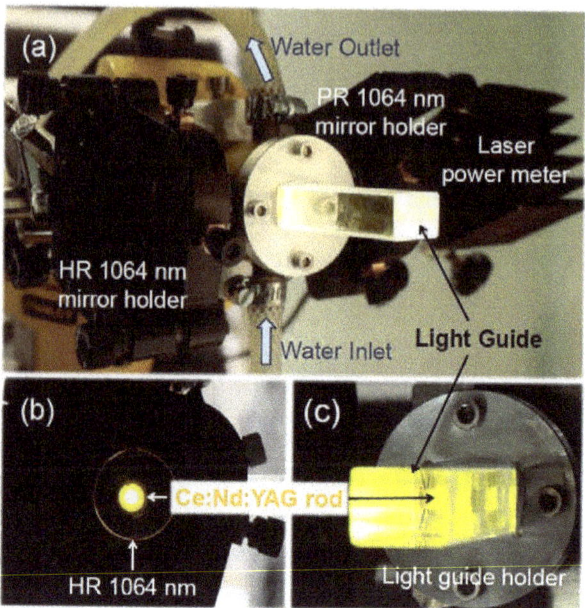

Figure 5. Photograph of (**a**) the Ce:Nd:YAG solar-pumped laser head with the laser resonator and (**b**,**c**) the detailed view of the Ce:Nd:YAG laser medium in the experiments. HR, high reflection; PR, partial reflection.

The present scheme was designed and produced in Lisbon. It was then tested at the 2 m diameter MSSF of the PROMES-CNRS, during the winter period of February 2022.

4. Solar-Pumped Ce:Nd:YAG Laser Experiments with the Rectangular Fused Silica Light Guide

To examine the Ce:Nd:YAG medium resistance under extreme solar pumping and, hence, its aptitude to scale to higher power levels, the performance of the Ce:Nd:YAG solar laser was evaluated as a function of the MSSF diameter D, starting from relatively small (D = 1.38 m) to the maximum diameter of the MSSF (D = 2.0 m). For maximum laser power extraction from the active medium in each case, the HR and PR laser mirrors were optically aligned as close as possible to the laser rod, as shown in Figure 5, forming a short and symmetric laser resonator with a 60 mm total length. Therefore, the Ce:Nd:YAG solar laser operated in a multimode regime. Flat laser mirrors were used to provide less laser beam divergence in relation to that with concave laser mirrors.

The influence of the incoming solar power on the Ce:Nd:YAG solar laser output power with four different Ds (1.38 m; 1.60 m; 1.78 m; 2.0 m) is given in Figure 6. During the experiments, the solar irradiances varied from 1000 to 1060 W/m^2, measured with a Kipp & Zonen CH1 pyrheliometer on a Kipp & Zonen 2AP solar tracker. Laser output power measurements were registered with a PM1100D power meter from Thorlabs.

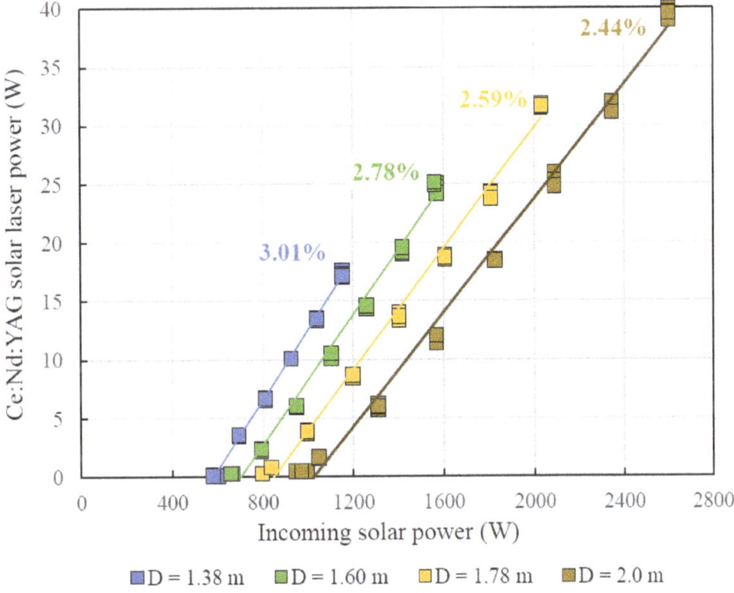

Figure 6. Variation of the Ce:Nd:YAG solar laser power with the incoming solar power and parabolic mirror diameter (D).

Table 1 summarizes the performance of the Ce:Nd:YAG solar laser output at each case, in terms of threshold solar power, maximum laser power, and slope efficiency. The M^2 beam quality factors for D = 1.38 m and D = 2.0 m are also given. Near-circular laser beam profiles and high evaporation rates were observed on an opaque material, placed at a distance of 1 m from the PR output mirror.

As observed in Figure 6, the variation of the solar concentrator diameter strongly influenced the laser performance of the Ce:Nd:YAG medium. It performed better with a lower D = 1.38 m, started to emit a laser at a minimum incoming solar power of 578 W, and reached a laser output power of 17.4 W at an incoming solar power of 1154 W. The slope efficiency of 3.01% was hence calculated in this case. $M^2_x \approx M^2_y \approx 38$ laser beam quality factors were also registered with D = 1.38 m. These results were slightly worse than that of the most recent Ce:Nd:YAG laser rod with a fused silica aspherical lens as a secondary concentrator, tested under similar pumping conditions, through which a 19.6 W cw laser power and 3.03% slope efficiency were measured [27]. The use of the aspheric lens helped to preserve the concentrated solar radiation profile, thus allowing to a certain extent a more effective side-pumping of the laser rod. However, at incoming solar powers greater than 1125 W, the non-uniformity of the absorption profile of the Ce:Nd:YAG laser rod led to its fracture [27], which means that the laser rod could not withstand more than 43% of the total incoming solar power of the MSSF. On the contrary, with the use of the light guide as a secondary concentrator, the Ce:Nd:YAG solar laser remained operational even with the maximum collection area of the MSSF, as shown in both Figure 6 and Table 1. Despite the decrease in slope efficiency to 2.44% and the deterioration of the laser beam quality to $M^2{}_x \approx 52$, $M^2{}_y \approx 54$ factors with an increasing collection area, a 40 W cw solar laser output power was reached for D = 2 m, without damaging the Ce:Nd:YAG laser rod. This was the highest laser power achieved with the Ce:Nd:YAG laser medium pumped by broadband solar radiation, being twice that obtained by the previous side-pumped Ce:Nd:YAG solar laser [27]. It was also 1.19 times more than the maximum solar laser power emitted from a side-pumped Cr:Nd:YAG laser rod using light guide [22].

Table 1. Ce:Nd:YAG solar laser performance with light guide with different parabolic mirror diameters (D).

Collection Diameter, D	D = 1.38 m	D = 1.60 m	D = 1.78 m	D = 2.00 m
Rim angle, α	44°	50°	55°	60°
Effective collection area	1.09 m^2	1.53 m^2	1.93 m^2	2.48 m^2
Maximum incoming solar power	1154 W	1559 W	2033 W	2600 W
Minimum threshold solar power	578 W	657 W	795 W	943 W
Maximum laser output power	17.35 W	25.13 W	31.80 W	40.01 W
Slope efficiency	3.01%	2.78%	2.59%	2.44%
$M^2{}_x$, $M^2{}_y$ factors	38, 38	—	—	52, 54

5. Conclusions

The potential of the Ce:Nd:YAG laser medium for solar laser power scaling was evaluated at the MSSF of PROMES-CNRS. The adoption of the side-pumping configuration with a rectangular light guide ensured a uniform pump light distribution along the laser rod. Thanks to this, the 4 mm diameter and 35 mm length Ce(0.1 at.%):Nd(1.1 at.%):YAG rod demonstrated a remarkable resistance to highly incoming solar powers, compared to the previous Ce:Nd:YAG solar lasers [25,27]. By using the total incoming solar power of the MSSF, a 40 W cw solar laser power was registered. As far as we are aware, this was the highest laser power level reported from a solar powered Ce:Nd:YAG laser medium, being two times more than that from the most recent Ce:Nd:YAG laser rod side-pumped with an aspheric lens secondary concentrator at the MSSF [27]. It was also the highest side-pumped solar laser power through light guide [20–22]. Therefore, the side-pumping of the laser rod, with the help of the fused silica light homogenizer, proved to be a good solution for scaling the solar laser power with the Ce:Nd:YAG laser medium.

In future work, either side-pumping or end-side-pumping configurations with light-guides could be employed for the simultaneous pumping of several Ce:Nd:YAG laser rods of smaller diameters within a common pumping cavity. The highly concentrated solar radiation could hence be evenly shared by the several laser rods, ensuring not only a substantial alleviation of the thermal lensing effects in solar-powered lasers [19], but also a significant rise in solar laser conversion efficiency. This may pave the way of Ce:Nd:YAG solar laser research into a new phase of development in terms of efficiency, laser beam quality, and stability at higher power levels.

Author Contributions: Conceptualization, J.A. and D.L.; methodology, J.A. and D.L.; validation, J.A., D.L. and C.R.V.; formal analysis, J.A., D.L. and C.R.V.; investigation, J.A., D.L., C.R.V., D.G. and E.G.; resources, D.L., J.A. and E.G.; data curation, J.A., C.R.V., D.L., D.G., B.D.T., H.C. and M.C.; writing—original draft preparation, J.A.; writing—review and editing, J.A., D.L., D.G., B.D.T., H.C.,

M.C. and C.R.V.; supervision, D.L.; project administration, D.L.; funding acquisition, D.L., J.A. and E.G. All authors have read and agreed to the published version of the manuscript.

Funding: This research was financially supported by Science and Technology Foundation of Portuguese Ministry of Science, Technology and Higher Education (FCT-MCTES) in the framework of the strategic project UIDB/00068/2020 and the exploratory research project EXPL/FIS-OTI/0332/2021. The solar laser research was also supported by the Solar Facilities for European Research Area–Third Phase (SFERA III), Grant Agreement No. 823802.

Institutional Review Board Statement: Not applicable.

Informed Consent Statement: Not applicable.

Data Availability Statement: Not applicable.

Acknowledgments: The authors express their gratitude for the FCT-MCTES fellowship grants CEECIND/03081/2017, PD/BD/142827/2018, PD/BD/128267/2016, 2021.06172.BD, SFRH/BD/145322/2019 and SFRH/BPD/125116/2016.

Conflicts of Interest: The authors declare no conflict of interest.

References

1. The Global Goals, Goal 7: Affordable and Clean Energy. Available online: https://www.globalgoals.org/goals/7-affordable-and-clean-energy/ (accessed on 1 April 2022).
2. The Global Goals, Goal 9: Industry, Innovation and Infrastructure. Available online: https://www.globalgoals.org/goals/9-industry-innovation-and-infrastructure/ (accessed on 1 April 2022).
3. Yabe, T.; Bagheri, B.; Ohkubo, T.; Uchida, S.; Yoshida, K.; Funatsu, T.; Oishi, T.; Daito, K.; Ishioka, M.; Yasunaga, N.; et al. 100 W-class solar pumped laser for sustainable magnesium-hydrogen energy cycle. *J. Appl. Phys.* **2008**, *104*, 083104. [CrossRef]
4. Motohiro, T.; Takeda, Y.; Ito, H.; Hasegawa, K.; Ikesue, A.; Ichikawa, T.; Higuchi, K.; Ichiki, A.; Mizuno, S.; Ito, T.; et al. Concept of the solar-pumped laser-photovoltaics combined system and its application to laser beam power feeding to electric vehicles. *Jpn. J. Appl. Phys.* **2017**, *56*, 08MA07. [CrossRef]
5. Abdel-Hadi, Y.A. Space-based solar laser system simulation to transfer power onto the earth. *NRIAG J. Astron. Geophys.* **2020**, *9*, 558–562. [CrossRef]
6. Weaver, W.R.; Lee, J.H. A Solar Pumped Gas Laser for the Direct Conversion of Solar Energy. *J. Energy* **1983**, *7*, 498–501. [CrossRef]
7. Kiss, Z.J.; Lewis, H.R.; Duncan, R.C. Sun pumped continuous optical maser. *Appl. Phys. Lett.* **1963**, *2*, 93–94. [CrossRef]
8. Insuik, R.J.; Christiansen, W.H. Blackbody-pumped CO_2 laser experiment. *AIAA J.* **1984**, *22*, 1271–1274. [CrossRef]
9. Terry, C.K.; Peterson, J.E.; Goswami, D.Y. Terrestrial solar-pumped iodine gas laser with minimum threshold concentration requirements. *J. Thermophys. Heat Transf.* **1996**, *10*, 54–59. [CrossRef]
10. Lee, J.H.; Kim, K.C.; Kim, K.H. Threshold pump power of a solar-pumped dye laser. *Appl. Phys. Lett.* **1988**, *53*, 2021–2022. [CrossRef]
11. Arashi, H.; Oka, Y.; Sasahara, N.; Kaimai, A.; Ishigame, M. A Solar-Pumped cw 18 W Nd:YAG Laser. *Jpn. J. Appl. Phys.* **1984**, *23*, 1051–1053. [CrossRef]
12. Weksler, M.; Shwartz, J. Solar-pumped solid-state lasers. *IEEE J. Quantum Electron.* **1988**, *24*, 1222–1228. [CrossRef]
13. Lando, M.; Kagan, J.; Linyekin, B.; Dobrusin, V. A solar-pumped Nd:YAG laser in the high collection efficiency regime. *Opt. Commun.* **2003**, *222*, 371–381. [CrossRef]
14. Dinh, T.H.; Ohkubo, T.; Yabe, T.; Kuboyama, H.J.O.L. 120 watt continuous wave solar-pumped laser with a liquid light-guide lens and an Nd:YAG rod. *Opt. Lett.* **2012**, *37*, 2670–2672. [CrossRef] [PubMed]
15. Liang, D.; Almeida, J. Highly efficient solar-pumped Nd:YAG laser. *Opt. Express* **2011**, *19*, 26399–26405. [CrossRef] [PubMed]
16. Liang, D.; Almeida, J.; Vistas, C.R.; Guillot, E. Solar-pumped Nd:YAG laser with 31.5W/m^2 multimode and 7.9W/m^2 TEM$_{00}$-mode collection efficiencies. *Sol. Energy Mater. Sol. Cells* **2017**, *159*, 435–439. [CrossRef]
17. Guan, Z.; Zhao, C.; Li, J.; He, D.; Zhang, H. 32.1 W/m^2 continuous wave solar-pumped laser with a bonding Nd:YAG/YAG rod and a Fresnel lens. *Opt. Laser Technol.* **2018**, *107*, 158–161. [CrossRef]
18. Guan, Z.; Zhao, C.; Zhang, H.; Li, J.; He, D.; Almeida, J.; Vistas, C.R.; Liang, D. 5.04% system slope efficiency solar-pumped Nd:YAG laser by a heliostat-parabolic mirror system. *J. Photonics Energy* **2018**, *8*, 2. [CrossRef]
19. Liang, D.; Almeida, J.; Garcia, D.; Tibúrcio, B.D.; Guillot, E.; Vistas, C.R. Simultaneous solar laser emissions from three Nd:YAG rods within a single pump cavity. *Sol. Energy* **2020**, *199*, 192–197. [CrossRef]
20. Almeida, J.; Liang, D.; Guillot, E. Improvement in solar-pumped Nd:YAG laser beam brightness. *Opt. Laser Technol.* **2012**, *44*, 2115–2119. [CrossRef]
21. Almeida, J.; Liang, D.; Vistas, C.R.; Bouadjemine, R.; Guillot, E. 5.5 W continuous-wave TEM$_{00}$-mode Nd:YAG solar laser by a light-guide/2V-shaped pump cavity. *Appl. Phys. B* **2015**, *121*, 473–482. [CrossRef]

22. Liang, D.; Almeida, J.; Guillot, E. Side-pumped continuous-wave Cr:Nd:YAG ceramic solar laser. *Appl. Phys. B Lasers Opt.* **2013**, *111*, 305–311. [CrossRef]

23. Yabe, T.; Ohkubo, T.; Uchida, S.; Yoshida, K.; Nakatsuka, M.; Funatsu, T.; Mabuti, A.; Oyama, A.; Nakagawa, K.; Oishi, T.; et al. High-efficiency and economical solar-energy-pumped laser with Fresnel lens and chromium codoped laser medium. *Appl. Phys. Lett.* **2007**, *90*, 261120. [CrossRef]

24. Liang, D.; Vistas, C.R.; Tibúrcio, B.D.; Almeida, J. Solar-pumped Cr:Nd:YAG ceramic laser with 6.7% slope efficiency. *Sol. Energy Mater. Sol. Cells* **2018**, *185*, 75–79. [CrossRef]

25. Vistas, C.R.; Liang, D.; Garcia, D.; Almeida, J.; Tibúrcio, B.D.; Guillot, E. Ce:Nd:YAG continuous-wave solar-pumped laser. *Optik* **2020**, *207*, 163795. [CrossRef]

26. Vistas, C.; Liang, D.; Almeida, J.; Tibúrcio, B.; Garcia, D.; Catela, M.; Costa, H.; Guillot, E. Ce:Nd:YAG side-pumped solar laser. *J. Photonics Energy* **2021**, *11*, 018001. [CrossRef]

27. Vistas, C.R.; Liang, D.; Garcia, D.; Catela, M.; Tibúrcio, B.D.; Costa, H.; Guillot, E.; Almeida, J. Uniform and Non-Uniform Pumping Effect on Ce:Nd:YAG Side-Pumped Solar Laser Output Performance. *Energies* **2022**, *15*, 3577. [CrossRef]

28. Lupei, V.; Lupei, A.; Gheorghe, C.; Ikesue, A. Emission sensitization processes involving Nd^{3+} in YAG. *J. Lumin.* **2016**, *170*, 594–601. [CrossRef]

29. Saiki, T.; Uchida, S.; Imasaki, K.; Motokoshi, S.; Yamanaka, C.; Fujita, H.; Nakatsuka, M.; Izawa, Y. Oscillation Property of Disk-Type Nd/Cr:YAG Ceramic Lasers with Quasi-Solar Pumping, (CLEO). In Proceedings of the Lasers and Electro-Optics, Baltimore, MD, USA, 22–27 May 2005; Volume 2, pp. 915–917.

30. Yagi, H.; Yanagitani, T.; Yoshida, H.; Nakatsuka, M.; Ueda, K. The optical properties and laser characteristics of Cr^{3+} and Nd^{3+} co-doped $Y_3Al_5O_{12}$ ceramics. *Opt. Laser Technol.* **2007**, *39*, 1295–1300. [CrossRef]

31. Payziyev, S.; Makhmudov, K.; Abdel-Hadi, Y.A. Simulation of a new solar Ce:Nd:YAG laser system. *Optik* **2018**, *156*, 891–895. [CrossRef]

32. Payziyev, S.; Sherniyozov, A.; Bakhramov, S.; Zikrillayev, K.; Khalikov, G.; Makhmudov, K.; Ismailov, M.; Payziyeva, D. Luminescence sensitization properties of Ce:Nd:YAG materials for solar pumped lasers. *Opt. Commun.* **2021**, *499*, 127283. [CrossRef]

33. Mares, J.; Jacquier, B.; Pédrini, C.; Boulon, G.J.R.P.A. Energy transfer mechanisms between Ce^{3+} and Nd^{3+} in YAG: Nd, Ce at low temperature. *Rev. Phys. Appliquée* **1987**, *22*, 145–152. [CrossRef]

34. Meng, J.X.; Li, J.Q.; Shi, Z.P.; Cheah, K.W. Efficient energy transfer for Ce to Nd in Nd/Ce codoped yttrium aluminum garnet. *Appl. Phys. Lett.* **2008**, *93*, 221908. [CrossRef]

35. Tai, Y.; Zheng, G.; Wang, H.; Bai, J. Near-infrared quantum cutting of Ce^{3+}–Nd^{3+} co-doped $Y_3Al_5O_{12}$ crystal for crystalline silicon solar cells. *J. Photochem. Photobiol. A Chem.* **2015**, *303–304*, 80–85. [CrossRef]

36. Guo, Y.; Huang, J.; Ke, G.; Ma, Y.; Quan, J.; Yi, G. Growth and optical properties of the Nd,Ce:YAG laser crystal. *J. Lumin.* **2021**, *236*, 118134. [CrossRef]

37. Villars, B.; Steven Hill, E.; Durfee, C.G. Design and development of a high-power LED-pumped Ce:Nd:YAG laser. *Opt. Lett.* **2015**, *40*, 3049–3052. [CrossRef] [PubMed]

38. Samuel, P.; Yanagitani, T.; Yagi, H.; Nakao, H.; Ueda, K.I.; Babu, S.M. Efficient energy transfer between Ce^{3+} and Nd^{3+} in cerium codoped Nd: YAG laser quality transparent ceramics. *J. Alloys Compd.* **2010**, *507*, 475–478. [CrossRef]

39. *ASTM G173-03*; Standard Tables for Reference Solar Spectral Irradiances: Direct Normal and Hemispherical on 37° Tilted Surface. ASTM International: West Conshohocken, PA, USA, 2012.

40. Nishiura, S.; Tanabe, S.; Fujioka, K.; Fujimoto, Y. Properties of transparent Ce: YAG ceramic phosphors for white LED. *Opt. Mater.* **2011**, *33*, 688–691. [CrossRef]

41. Prahl, S. Nd:YAG—Nd:Y3Al5O12. Available online: https://omlc.org/spectra/lasermedia/html/052.html (accessed on 1 April 2022).

42. Liu, X.; Teng, Y.; Zhuang, Y.; Xie, J.; Qiao, Y.; Dong, G.; Chen, D.; Qiu, J. Broadband conversion of visible light to near-infrared emission by Ce^{3+}, Yb^{3+}-codoped yttrium aluminum garnet. *Opt. Lett.* **2009**, *34*, 3565–3567. [CrossRef]

43. PROMES-CNRS Solar Furnaces and Concentrating Solar Systems. Available online: https://www.promes.cnrs.fr/en/infrastructure-solaire/moyens-solaires/solar-furnaces-and-concentrating-solar-systems/ (accessed on 1 April 2022).

44. Heraeus Properties of Fused Silica. Available online: https://www.heraeus.com/en/hca/fused_silica_quartz_knowledge_base_1/properties_1/properties_hca.html (accessed on 1 April 2022).

45. Mehellou, S.; Liang, D.; Almeida, J.; Bouadjemine, R.; Vistas, C.R.; Guillot, E.; Rehouma, F. Stable solar-pumped TEM_{00}-mode 1064 nm laser emission by a monolithic fused silica twisted light guide. *Sol. Energy* **2017**, *155*, 1059–1071. [CrossRef]

Article

Ce:Nd:YAG Solar Laser with 4.5% Solar-to-Laser Conversion Efficiency

Dário Garcia, Dawei Liang *, Cláudia R. Vistas, Hugo Costa, Miguel Catela, Bruno D. Tibúrcio and Joana Almeida

Centre of Physics and Technological Research, Departamento de Física, Faculdade de Ciências e Tecnologia, Universidade NOVA de Lisboa, 2829-516 Caparica, Portugal; kongming.dario@gmail.com (D.G.); c.vistas@fct.unl.pt (C.R.V.); hf.costa@campus.fct.unl.pt (H.C.); m.catela@campus.fct.unl.pt (M.C.); brunotiburcio78@gmail.com (B.D.T.); jla@fct.unl.pt (J.A.)
* Correspondence: dl@fct.unl.pt

Abstract: The efficiency potential of a small-size solar-pumped laser is studied here. The solar laser head was composed of a fused silica aspheric lens and a conical pump cavity, which coupled and redistributed the concentrated solar radiation from the focal zone of a parabolic mirror with an effective collection area of 0.293 m^2 to end-side pump a Ce (0.1 at%):Nd (1.1 at%):YAG rod of 2.5 mm diameter and 25 mm length. Optimum solar laser design parameters were found through Zemax$^©$ non-sequential ray-tracing and LASCAD™ analysis. The utilization of the Ce:Nd:YAG medium with small diameter pumped by a small-scale solar concentrator was essential to significantly enhance the end-side pump solar laser efficiency and thermal performance. For 249 W incoming solar power at an irradiance of 850 W/m^2, 11.2 W multimode solar laser power was measured, corresponding to the record solar-to-laser power conversion efficiency of 4.50%, being, to the best of our knowledge, 1.22 times higher than the previous record. Moreover, the highest solar laser collection efficiency of 38.22 W/m^2 and slope efficiency of 6.8% were obtained, which are 1.18 and 1.02 times, respectively, higher than the previous records. The lowest threshold solar power of a Ce:Nd:YAG solar-pumped laser is also reported here.

Keywords: solar-pumped laser; Ce:Nd:YAG; parabolic mirror; laser efficiency

Citation: Garcia, D.; Liang, D.; Vistas, C.R.; Costa, H.; Catela, M.; Tibúrcio, B.D.; Almeida, J. Ce:Nd:YAG Solar Laser with 4.5% Solar-to-Laser Conversion Efficiency. *Energies* **2022**, *15*, 5292. https://doi.org/10.3390/en15145292

Academic Editor: Tapas Mallick

Received: 29 June 2022
Accepted: 19 July 2022
Published: 21 July 2022

Publisher's Note: MDPI stays neutral with regard to jurisdictional claims in published maps and institutional affiliations.

1. Introduction

Solar-pumped lasers can directly convert the incoherent broadband radiation into coherent narrowband radiation through an active medium, while the conventional lasers utilize the electric-to-light conversion from diodes or lamps. The direct harnessing of solar energy is a promising renewable technology that may bring important economic advantages for countries with high solar availability [1], for the development of sustainable industry [2], on earth [3,4] or in space [5].

In 1963, a calcium fluoride crystal doped with divalent dysprosium (Dy^{2+}:CaF$_2$) was used to generate the first continuous-wave laser emission of 2.36 μm, by Kiss et al. [6]. From then on, several attempts to increase the solar laser efficiency have been made using different gain medium, such as gas [7,8], liquids [9], and solids [10,11]. However, only solid active media was promising enough for solar laser emission, especially the yttrium aluminum garnet (Y$_3$Al$_5$O$_{12}$, YAG) as hosting crystal structure and the rare earth ion of neodymium (Nd^{3+}) as an activator ion, contributing to the significant advances in solar laser efficiency. About 18.7 W/m^2 solar laser collection efficiency, defined as laser output power versus solar collector area, was reported in 2007 by pumping a large Cr:Nd:YAG ceramic laser rod with a 1.4 m^2 Fresnel lens primary concentrator [12]. 19.3 W/m^2 laser collection efficiency was later achieved in 2011 by exciting a 4 mm diameter, 25 mm length Nd:YAG single-crystal rod through a 0.64 m^2 area Fresnel lens [13]. In 2012, the Nd:YAG solar laser collection efficiency was boosted to 30.0 W/m^2 by pumping a 6 mm diameter, 100 mm length Nd:YAG single-crystal rod through a 4.0 m^2 area Fresnel lens [14]. According to the study of Zhao et al. [15], the Nd^{3+} ion absorption bands have about a 16% overlap with the solar spectrum. The solar laser output can be enhanced by adding sensitizer elements, such as chromium (Cr^{3+}) or cerium (Ce^{3+}) ions, which have broader absorptions bands in the visible region, increasing the energy conversion efficiency [16,17]. Although both Cr^{3+} and Ce^{3+} ions have different emission bands than that of the Nd^{3+} ion, they overlap perfectly to some of the Nd^{3+} absorption bands. Solar-pumped lasers with a Cr:Nd:YAG ceramic rod have been carried out since 2007, first by Yabe et al. [12] and later by Liang et al. in 2013 [18] and 2018 [19], but the results were not very significant, being only slightly higher than that with the Nd:YAG medium [20]. Initial stages of solar-pumping the Ce:Nd:YAG medium were conducted in the early 2020s [20], with a 5 mm diameter, 30 mm length Ce (0.05 at%): Nd (1.0 at%): YAG rod. It was end-side pumped by 960 W of incoming solar power, provided by a medium size solar furnace (MSSF) with 1.23 m^2 effective collection area in PROMES-CNRS [21]. In these conditions, 6.0 W laser output power was obtained. However, when increasing the incoming solar power, the rod got fractured in the upper-end region. A side-pumping scheme, which helps to spread the solar pumping radiation along the laser rod, was then adopted and tested at the NOVA heliostat-parabolic mirror system [22]. A successful solar laser emission of 16.5 W was achieved under an incoming solar power of 600 W with a Ce (0.1 at%):Nd (1.1 at%):YAG laser rod. This laser had 23.6 W/m^2 collection efficiency and 2.8% solar-to-laser conversion efficiency, with the latter being 1.57 times higher than that of a Nd (1.0 at%):YAG laser rod under the same conditions [22].

In 2022, Vistas et al. further explored the influence of the solar pumping distribution on a 4 mm diameter, 35 mm length Ce (0.1 at%):Nd (1.1 at%):YAG laser rod by testing two different secondary concentrators [23]: a point-focusing fused silica aspheric lens and a fused silica light guide that provided a homogenous light distribution at its output. With collection area of 1.09 m^2 from the MSSF parabolic mirror, the solar pumping scheme with the aspheric lens produced up to 19.6 W solar laser power before the fracture of the rod due to high thermal load, while the light guide scheme maintained continuous lasing operation, producing 17.4 W solar laser power at an incoming solar power of 1154 W. This resulted in 18 W/m^2 solar laser collection efficiency and 1.7% solar-to-laser conversion efficiency. The Ce:Nd:YAG solar laser initiated at an incoming solar power of 578 W, leading to 3.0% slope efficiency [23]. In the same year, Almeida et al. [24] explored the laser power scaling potential of side-pumped Ce:Nd:YAG with light guide using 2.48 m^2 effective collection area of the MSSF parabolic mirror [24]. 40 W solar laser output was achieved without fracturing the crystal. However, the collection, conversion, and slope efficiencies were reduced to 16.13 W/m^2, 1.54%, and 2.44% slope efficiencies, respectively, and the threshold power increased to 943 W. This demonstrates that the efficiency of solar laser output tends to be lower with larger collection area.

Due to the above mentioned reasons, the solar pumping of a thin Ce:Nd:YAG rod through a primary concentrator with a small effective collection area was explored here. A fused silica aspheric lens was used as a secondary concentrator for end-side-pumping the 2.5 mm diameter, 25 mm length Ce (0.1 at%):Nd (1.1 at%):YAG rod through the NOVA parabolic mirror with 0.293 m^2 effective collection area. During the experiments, 850 W/m^2 solar irradiance was measured, leading to an incoming solar power of 249.05 W. This constituted a substantial reduction from the 600 W incoming power of the most efficient Ce:Nd:YAG solar laser [22] at the same solar facility. Moreover, solar laser emission was possible, unlike the previous end-side-pumping [20] and side-pumping [23] Ce:Nd:YAG solar lasers in which the rod was fractured due to high concentrated local heat load. This resulted in 11.2 W solar laser power, corresponding to a solar laser collection efficiency of 38.22 W/m^2 and slope efficiency of 6.8%, which are 1.18 and 1.02 times higher than the previous state-of-the-art record [19]. More importantly, 4.50% solar-to-laser power conversion efficiency was achieved, also representing an enhancement of 1.22 times [19]. Moreover, only 88 W incoming solar power was required for Ce (0.1 at%): Nd(1.1 at%):YAG to start lasing, as compared to the minimum of 200 W for Nd:YAG [25]. Moreover, the substantial reduction of the collection area from 1.03 m^2 [25] to 0.293 m^2 may motivate the research of small size solar laser models in direct solar tracking mode with promising possibility of further solar laser efficiency enhancement.

2. Materials and Method

2.1. Solar Energy Collection and Concentration: NOVA Heliostat-Parabolic System

The NOVA heliostat-parabolic mirror solar laser system, as shown in Figure 1, is composed of a two-axis heliostat with four flat mirrors of 93.5% reflectivity, which are used for redirecting the incoming solar radiation toward a parabolic mirror of 80.0% reflectivity with a focal length of 660 mm. This whole system has a combined reflective capacity of 74.8%. By using an external annular mask, the collection diameter of the parabolic mirror was reduced from 1.50 m to 0.68 m. About 0.293 m^2 effective solar collection area was then calculated by discounting the shadowed area of approximately 0.07 m^2 caused by the laser head and its supporting mechanics, the *X-Y-Z* positioner, and the non-reflecting space between the two flat segments of the heliostat.

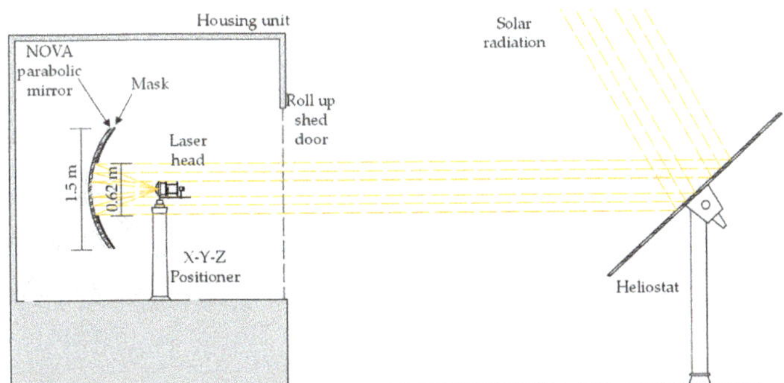

Figure 1. Schematic of the NOVA heliostat-parabolic mirror solar energy collection and concentration system.

2.2. Solar Laser Head

Figure 2 shows the image of the solar laser head and its optical components, as well as the laser resonator components. The laser head was composed of a large fused silica aspheric lens of 99.995% optical purity that focused the concentrated solar radiation onto the flat upper surface of the Ce (0.1 at%):Nd (1.1 at%):YAG rod of 2.5 mm diameter and 25 mm length, mounted within a single reflective conical pump cavity. The rod's flat upper surface had a highly reflective (HR) coating for the laser emission wavelength at 1064 nm (99.9% @ 1064 nm), while the other end surface had an anti-reflection (AR) coating for the laser emission wavelength (reflectivity (R) < 0.2% @ 1064 nm), supplied by Chengdu, Dongjun Laser Co., Ltd. (Chengdu, China). Cooling water constantly flowed into the laser head from an inlet and the heated water was extracted through the outlet. Maximum lasing could only be achieved when the laser head was correctly aligned at the focus of solar concentrator through the *X-Y-Z* positioner and the correct alignment of the laser resonator mirrors.

The fused silica aspheric lens had an 82 mm diameter and a thickness of 37 mm, with a curved input surface and a flat output surface. The curved feature of the aspheric lens follows the sag (z) in Equation (1), with radial aperture (r) of 41 mm, the parabolic constant $k = 0$, the radius of curvature (c) of −43 mm, and the aspheric coefficient $\beta_1 = -0.004$.

$$z = \frac{c \times r^2}{1 + \sqrt{1 - (1+k)c^2 r^2}} + \beta_1 r^2, \tag{1}$$

The flat output surface of the lens was 6 mm apart from the flat input surface of the crystal rod. This space was necessary for the laser rod to effectively absorb the concentrated solar radiation and to guarantee an abundant cooling for effective heat dissipation. The absorption efficiency was further enhanced by a silver-coated aluminum foil of 94% reflectivity that covered the conical pump cavity. The reflective conical surface allowed the reabsorption of the solar energy due to the crisscross of the solar rays into the laser rod, thus helping in the redistribution of the solar energy along the rod. This conical pump cavity was 19.5 mm in length with input and output aperture diameters of 18 mm and 9 mm, respectively.

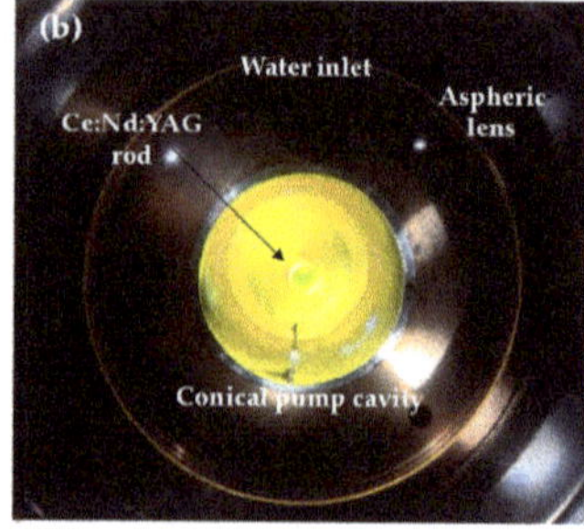

Figure 2. Photograph of (**a**) the Ce (0.1 at%):Nd (1.1 at%):YAG solar laser head at the focus of the parabolic mirror, and (**b**) the front view of the laser head.

Cooling water with 6 L/min flow rate was used as the heat extraction medium. The cooling system was designed to maximize the removal of the heat generated within the rod and the optical components, such as the fused silica aspheric lens and the conical pump cavity. Both the cooling water and the silica lens weaken the intensity of the UV solarization and IR heating arriving at the rod, which help to reduce the thermal lensing issue.

3. Theory

Energy Transfer Mechanism between Ce^{3+} and Nd^{3+} Ions in YAG

The study of Ce^{3+} ion doped in Nd:YAG under broadband pumping has been a topic of research for decades, starting in 1969 by Holloway et al. [26]. In 1987, Mares et al. studied the energy transfer mechanism between Ce^{3+} and Nd^{3+} in YAG at low temperatures [27]. Later in 2015, Tai et al. explored the mechanism of energy transfer of Ce^{3+} to Nd^{3+} in YAG with quantum cutting [28]. Its recent use in 2021 by Vistas et al., as a substitute of Nd:YAG in solar-pumped lasers, has shown promising results, achieving about 1.6 times more collection efficiency and solar-to-laser power conversion efficiency compared to that with Nd:YAG crystal rod under the same conditions [22].

The direct solar spectrum, the Ce:Nd:YAG absorption spectrum, the Ce^{3+} and Nd^{3+} ions' emission bands in YAG are presented in Figure 3. The broad absorption bands of the Ce^{3+} ion is located at the most energetic region of the electromagnetic spectrum, between 315 nm and 510 nm, corresponding to the excited energy level of $5d_2$ (centered at 339 nm) and $5d_1$ (centered at 460 nm) [28]. While the Nd^{3+} absorption bands are in between 510 nm and 888 nm, with five prominent absorbing bands, $^4G_{7/2} + {}^2G_{9/2} + {}^2K_{13/2}$ (centered at 530 nm), $^2G_{7/2} + {}^4G_{5/2}$ (centered at 580 nm), $^4S_{3/2} + {}^4F_{7/2}$ (centered at 770 nm), $^2H_{9/2} + {}^4F_{5/2}$ (centered at 805 nm), and $^4F_{3/2}$ (centered at 878 nm) [29].

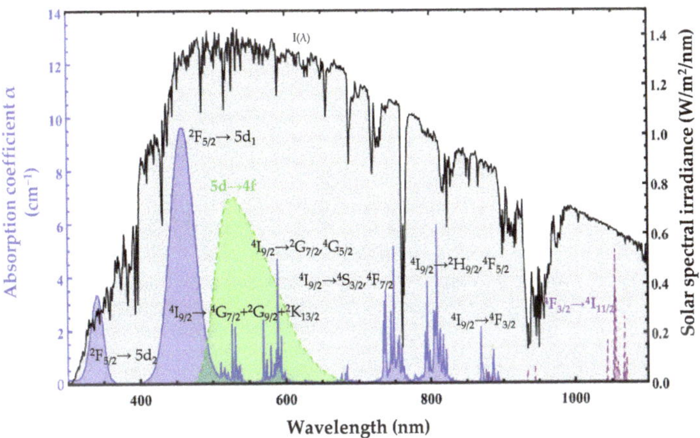

Figure 3. AM1.5 direct solar spectrum (black), adapted from [30]. Ce (0.1 at%):Nd (1.1 at%):YAG absorption spectrum (blue), Ce^{3+} (green) and Nd^{3+} (purple) emission spectra in YAG, adapted from [31].

The strong UV and blue photons excite the Ce^{3+} 4f level (the $^2F_{5/2}$ ground state), toward the 5d sublevels of $5d_1$ and $5d_2$, as shown in Figure 4 [28,32]. The consequent energy relaxation from $5d_1$ to both ground states $^2F_{7/2}$ and $^2F_{5/2}$ forms a yellow/green broad emission spectrum, overlapping with the excitation peaks of the Nd^{3+} ion at the $^4G_{7/2} + {}^2G_{9/2} + {}^2K_{13/2}$ energy levels between 510 nm and 540 nm and the $^2G_{7/2} + {}^4G_{5/2}$ energies levels between 566 nm and 595 nm [28,33]. The radiative transfer mechanism by the transition 5d→4f of the Ce^{3+} ion and the transitions $^4I_{9/2} \rightarrow {}^4G_{7/2} + {}^2G_{9/2} + {}^2K_{13/2}$ and $^4I_{9/2} \rightarrow {}^2G_{7/2} + {}^4G_{5/2}$ through cross-relaxation process are indicated by pathway (1) in Figure 4 [28]. The pathway (2) represents the energy transfer from the Ce^{3+} excited state at $5d_1$ to the two Nd^{3+} ions at $^4F_{3/2}$ level through quantum cutting down-conversion process [28].

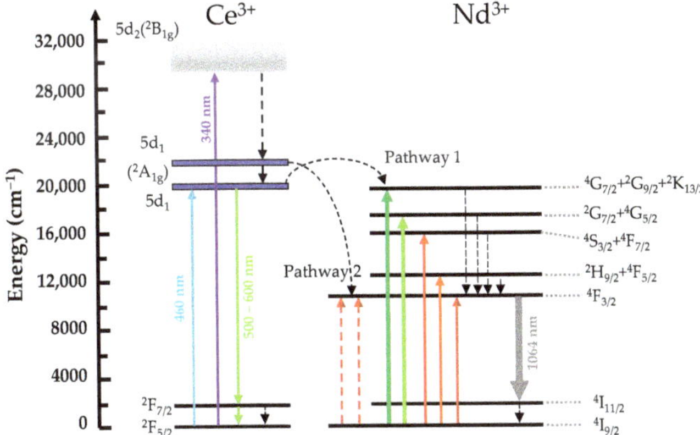

Figure 4. Absorption and emission energy level of Ce^{3+} and Nd^{3+} ions. (1) The energy transfer by cross relaxation of Ce^{3+} to Nd^{3+} ion, and (2) the quantum cutting down conversion of Ce^{3+} ion from 5d1 energy level into two Nd^{3+} ions at $^4F_{3/2}$ energy level. Adapted from [28,29].

The total absorbed solar power of an ion ($P_{abs,ion}$) of the active medium is calculated through Equation (2).

$$P_{abs,ion} = \int_{\lambda_i}^{\lambda_f} I(\lambda)\left(1 - e^{(-\alpha_{ion}(\lambda) \times L_c)}\right) d\lambda, \tag{2}$$

$I(\lambda)$ is the solar spectral irradiance (W/m^2/nm) of each wavelength λ, L_c is the effective absorption length of the laser rod and $\alpha_{ion}(\lambda)$ is the absorption coefficient of the ion for each wavelength. To effectively calculate the total power available from the Sun to feed both ions, L_c must be considered infinite. In this case, the total absorbed power by Ce^{3+} ($P_{abs,Ce}$) and Nd^{3+} ($P_{abs,Nd}$) ions is approximately 157 W and 162 W, respectively. Given the total solar irradiance of 1000 W/m^2, by integrating all the solar wavelength intensities, the overlap efficiency, i.e., the fraction of the available solar power that is absorbed by the Ce^{3+} ($\eta_{overlap,Ce}$) and Nd^{3+} ($\eta_{overlap,Nd}$) ions, is about 15.7% and 16.2%, respectively, as calculated through Equation (3).

$$\eta_{overlap,ion} = \frac{P_{abs,ion}}{\int I(\lambda)d\lambda}, \tag{3}$$

The energy transfer efficiency of the non-radiative process ($\eta_{NR:Ce\text{-}Nd}$) is about 70% [28,34], whereas for the radiative process ($\eta_{R:Ce\text{-}Nd}$) is about 30% [28].

4. Calculation

4.1. Zemax$^{©}$ Simulation

There are many possible and viable ways to calculate the absorbed power within the Ce:Nd:YAG with the prevalent wavelengths in the Zemax$^{©}$ software. A direct approach for calculation consists of using all the wavelengths of both the solar emission and the Ce:Nd:YAG absorption spectra. However, this method would require much more computing resources to achieve the same results as those by simulating with a few key solar wavelengths related to the absorption bands responsible for lasing [35].

A total available solar power of 249.05 W can be obtained by the NOVA parabolic mirror with an effective collection area (A) of 0.293 m^2, at 850 W/m^2 solar irradiance (I). However, only 16.2% and 15.7% of the total power calculated from Equation (3) are useful for calculating the absorbed power by the Ce:Nd:YAG rod with finite dimensions. Consequently, in Zemax$^{©}$, two sources were defined for solar pumping. The energy division is shown in Figure 5a.

Source 1 emits all the relevant overlapped wavelengths of the solar spectrum with the Nd^{3+} ion absorption spectrum, as well as the wavelength data that includes the Ce^{3+} quantum cutting down conversion of non-radiative transfer to Nd^{3+}, described by Pathway (1), as shown in Figure 5b. The simulated power of source 1 ($P_{source1}$) follows the Equation (4) and has a total power of 67.72 W.

$$P_{source\,1} = A\,I\left(\eta_{overlap,Nd^{3+}} + \eta_{overlap,Ce^{3+}} \times \eta_{NR:Ce-Nd}\right), \tag{4}$$

Source 2 accounts for 30% of the absorbed energy by Ce^{3+} that was transferred radiatively to Nd^{3+}, as shown in Figure 4 by Pathway (2). This source emits green and orange wavelengths useful for Nd^{3+} absorption, as shown in Figure 5c. The simulated power of source 2 ($P_{source2}$) follows the Equation (5) and has a total power of 11.73 W.

$$P_{source\,2} = A\,I\,\eta_{overlap,Ce^{3+}} \times \eta_{R:Ce-Nd}, \tag{5}$$

The amount of energy retained within the Ce (0.1 at%):Nd (1.1 at%):YAG depends on the transmission (T) data of the Nd^{3+} active ion, water and silica materials, shown in Equation (6).

$$T(\lambda) = e^{-\alpha(\lambda) \times L_c}, \tag{6}$$

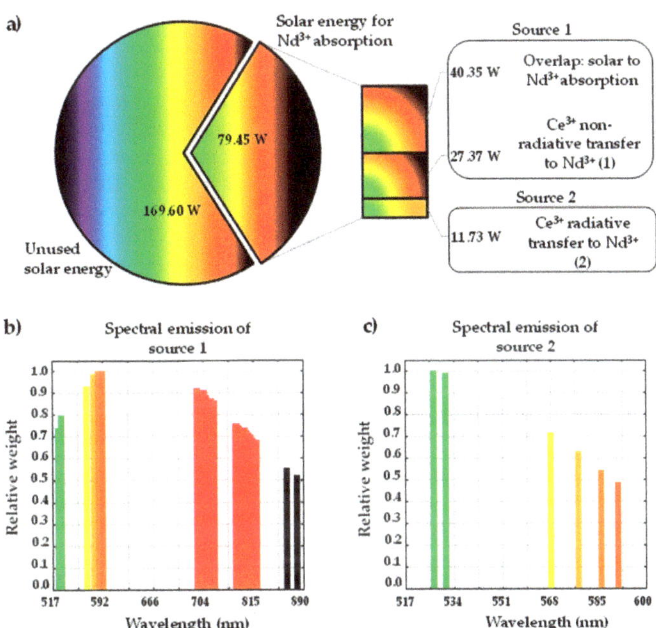

Figure 5. (**a**) Solar energy division for Ce:Nd:YAG absorption from a total of 249.05 W. Spectral composition used in (**b**) source 1 and (**c**) source 2.

Figure 6a shows the transmission data of water and fused silica materials from Zemax$^©$ glass catalog, as well as the 22 transmission wavelengths of Ce (0.1 at%):Nd (1.1 at%):YAG, at L_c = 10 mm. Figure 6b shows the index of refraction of Ce (0.1 at%):Nd (1.1 at%):YAG, water and fused silica as a function of the wavelength utilized in Zemax$^©$.

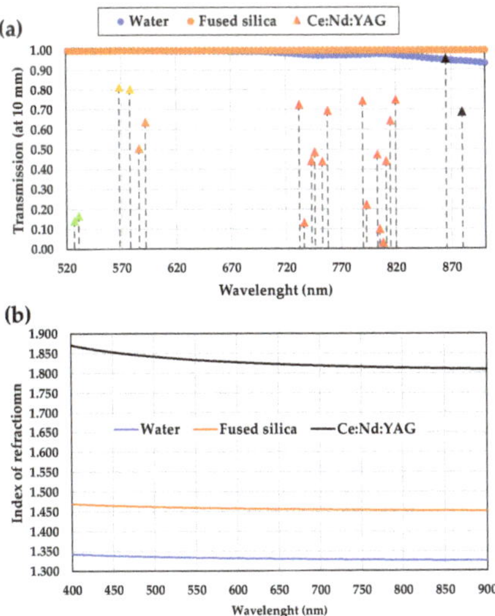

Figure 6. Materials in Zemax$^©$: Ce (0.1 at%):Nd (0.1 at%):YAG, water and fused silica. (**a**) Transmission of the materials at 10 mm depth. (**b**) Index of refraction of the materials at 20 °C and 1 atm.

The transmission and the refractive index of the optical materials, the shape of each optical apparatus, the solar wavelengths and the angle of the concentrated solar rays determine the optical path of the solar rays from the source to the crystal rod. Figure 7 shows the optical path of the concentrated solar rays arriving at the laser head, through the fused silica aspheric lens, then into the cooling water and finally to the Ce:Nd:YAG crystal rod. Most of the concentrated solar rays enter the laser rod through its upper HR1064 nm end face, and travel within the rod through total internal reflection (due to the refractive index differences between the water and the active medium), as shown in Figure 7 by the red ray. Side-pumping occurs when rays enter the rod through its lateral surface. Multiple passes can be achieved by a single ray zigzagging back into the crystal through the conical pump cavity, which increases the path length of the ray inside the rod, L_p, and hence the amount of energy absorbed.

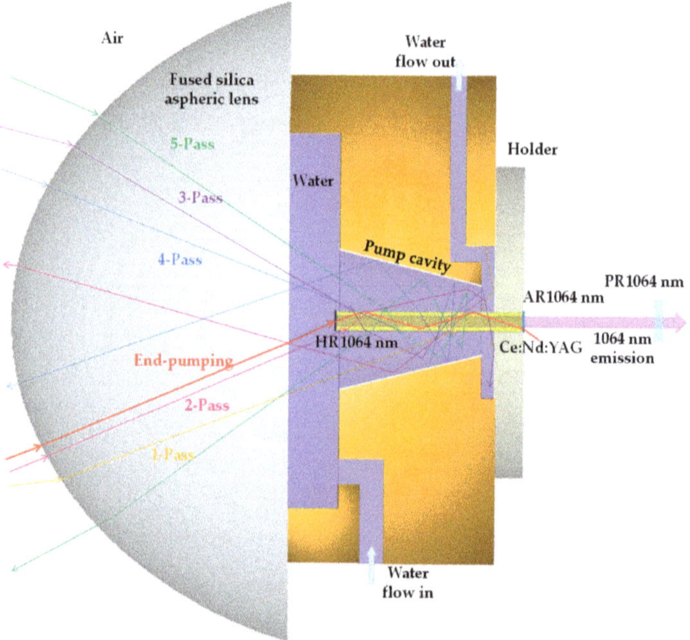

Figure 7. Solar rays at various angles and wavelengths passing through the Ce:Nd:YAG medium inside the laser head. End-pumping occurs through total internal reflection within the rod. Side-pumping may have 1 to 5 or more passes through the rod.

Figure 8 shows the absorption of the laser rod as a function of the wavelength and the number of passes. The end-pump has a minimum L_p of 25 mm, the same as the length of the rod. The average L_p of 1-pass, 2-pass, 3-pass, 4-pass, and end-pumping is about 3.25 mm, 6.50 mm, 9.75 mm, 13 mm, and 35 mm, respectively. The absorption of a single ray increases gradually with each successive number of passes within the crystal. Some wavelengths, such as 527 nm, 531 nm, 736 nm, 793 nm, 805 nm, and 808 nm are totally absorbed by the rod if the ray is traveling within the rod by end-pumping regime.

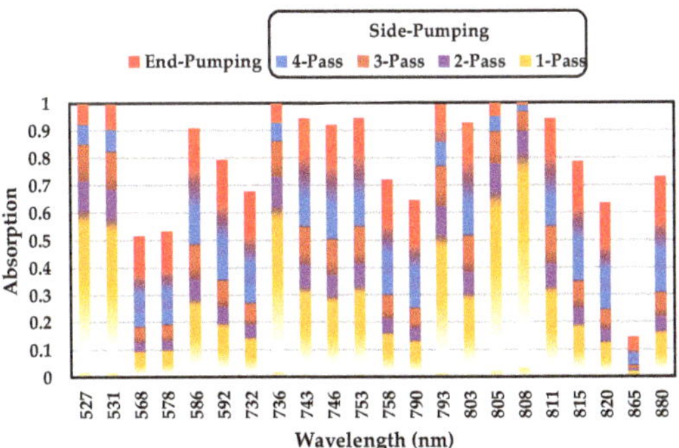

Figure 8. Wavelength-dependent absorption versus number of passes within the crystal. Each pass covers an average absorption length traveled within the crystal rod.

In Zemax$^©$, the Ce (0.1 at%):Nd (1.1 at%):YAG rod was divided into 125,000 voxels. The pump flux distribution along its longitudinal cross-section and five central transversal cross-sections are shown in Figure 9. Red color represents the maximum solar energy absorption, whereas blue represents little or no absorption. The most intensive absorbing region has a peak flux of 0.95 W/mm^3, located slightly below the input surface of the rod. The lower part of the rod also absorbs some radiation due to multiple passages of the solar rays by side-pumping with the help of the reflective cone. A total absorbed solar energy of 30.96 W was stored along the thermally loaded crystal.

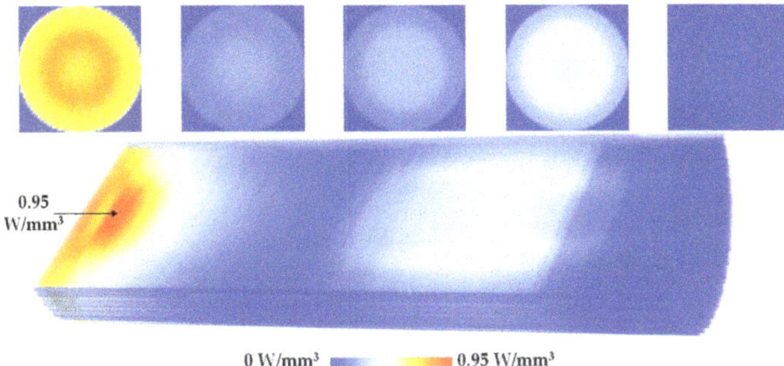

Figure 9. Absorbed solar pump-flux distribution along the longitudinal at a central cross-section, and five separate transversal cross-sections of the Ce:Nd:YAG rod.

4.2. LASCAD™ Simulation

The absorbed power in cubic matrix form was imported into the laser cavity analysis and design (LASCAD™) software to analyze the solar laser output performance. The material data of Ce (0.1 at%):Nd (1.1 at%):YAG crystal with a stimulated emission cross-section of 2.8×10^{-19} cm^2, a fluorescence lifetime of 230 μs and an absorption/scattering loss of 0.002 cm^{-1} was set in LASCAD™. The mean absorbed and intensity-weighted wavelength of 660 nm was considered in the analysis [11]. Figure 10 shows the laser resonator of the thermally loaded crystal rod and the associated dielectric lenses. The left end-face of the rod has a layer of HR 1064 mm coating, represented by the optical dielectric interface 0. The output end-face has an AR 1064 nm coating, represented by the optical dielectric interface 1. The partial reflection (PR) output mirror is positioned 17 mm apart from the AR coated end face (L_{AR-PR} = 17 mm), represented by optical dielectric interface 2. Different reflectivities and radii of curvature (RoC) of the PR mirror were tested to find the best combination to achieve the highest laser output power for the most efficiently solar-pumped Ce:Nd:YAG rod. A total round-trip loss of 1.6% was accounted in the LASCAD™ calculation.

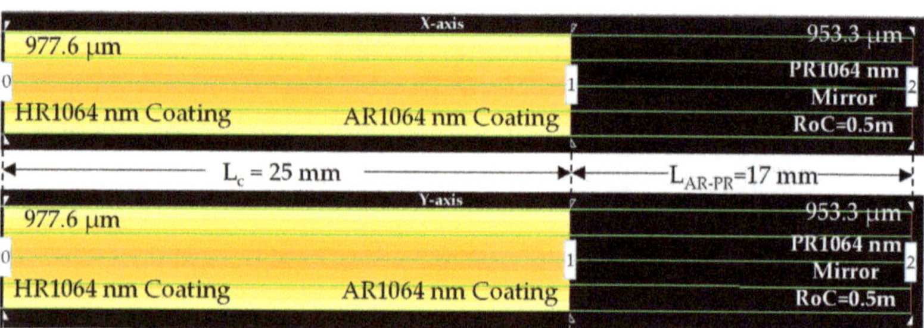

Figure 10. Laser resonator configuration for multimode solar laser extraction from the Ce:Nd:YAG rod. The L_c is the laser rod length, and the L_{AR-PR} is the length between AR1064 nm end face of the rod and the PR1064 nm mirror.

Figure 11 shows the heat load, the temperature, and the stress intensity of the 2.5 mm diameter, 25 mm length rod numerically calculated through LASCAD™ analysis, under ambient temperature and cooling boundary condition of 300 K. The maximum heat load of 0.385 W/mm^3 was locally found. A maximum rod temperature of 313.8 K is attained at the tip of the input surface of the crystal. A moderate maximum thermal stress of 24.23 N/mm^2 was found.

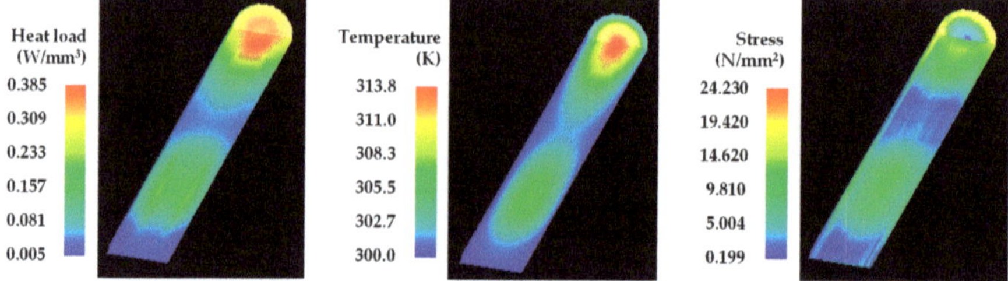

Figure 11. Heat load, temperature, and stress intensity of the 2.5 mm diameter, 25 mm length Ce:Nd:YAG rod, numerically acquired through LASCAD™ analysis.

During the solar laser experiments, the thin Ce:Nd:YAG rod was cooled by water with a constant flow rate of 6 L/min at 300 K ambient temperature. This allowed the maximum temperature within the crystal to rise no more than just 14 K with the concentrated solar pumping, as shown in Figure 11. According to [36], this temperature variation leads to a decrease in the stimulated emission cross section of only 0.05×10^{-19} cm^2 approximately. Hence, its effect on the Nd^{3+} ions absorption bands is minimal.

Considering all the mentioned physical characteristics of the stimulated active medium, the highest multimode laser power of 11.3 W was numerically calculated with a PR1064 nm of R = 96%, RoC of 0.5 m, as shown in Figure 12. The experimental results are also given.

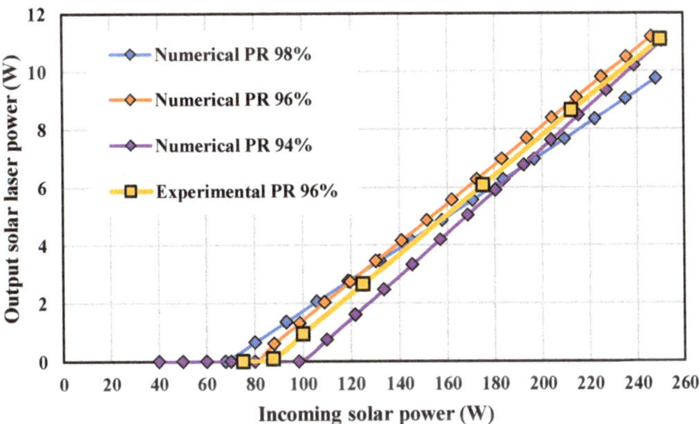

Figure 12. Ce:Nd:YAG solar laser output power as a function of the incoming solar power, numerically obtained using partial reflection (PR) mirror of 94%, 96%, and 98% and experimentally achieved with PR 96%.

5. Results and Discussion

The solar laser head prototype was built according to the model designed in Zemax$^©$ that provided the highest laser power extraction through LASCAD™. The laser head was tested at the NOVA solar facility in April of 2022. The solar irradiance of 850 W/m^2 was measured during that period, which is equivalent to a maximum incoming solar power of 249 W provided from the primary parabolic mirror with effective diameter of 0.68 m. A maximum multimode laser power of 11.2 W was successfully achieved using a PR (R $\geq$ 96% @ 1064 nm) mirror with RoC of 0.5 m, as shown in Figure 12, which matches well with the numerical prediction. The lasing emission started with an incoming solar power of 88 W and grew linearly to 11.2 W at 249 W incoming solar power, leading to a slope efficiency of 6.8%.

Table 1 summarizes the most recent progress in solar-pumped lasers, regarding to the minimum threshold power, maximum solar-to-laser power conversion, solar laser collection, and slope efficiencies.

In this work, only 88 W solar power was needed to start the lasing process, which is 0.44 times below the record minimum threshold power of 200 W [25]. The solar lasing system was able to produce 11.2 W multimode laser power, equivalent to the highest solar-to-laser power conversion efficiency of 4.50%, to the best of our knowledge, which is 1.22 times more than the record of 3.69% [19]. Furthermore, the solar laser collection and slope efficiencies were 38.22 W/m^2 and 6.8%, respectively, being 1.18 and 1.02 times higher than the previous records of 32.5 W/m^2 and 6.7% through an end-side-pumped Cr:Nd:YAG rod [19].

Table 1. Comparison of progress in solar efficiencies.

Parameters	Guan et al. 2018 [25]	Liang et al. 2018 [19]	Vistas et al. 2021 [22]	This Work 2022	Improvements Over Previous Record (Times)
Primary concentrator	Fresnel lens	Parabolic mirror	Parabolic mirror	Parabolic mirror	-
Overall efficiency of the collection system	~45%	75%	75%	75%	-
Effective collection area	1.030 m^2	1.000 m^2	1.070 m^2	0.293 m^2	-
Tracking method	Direct tracking	Via heliostat	Via heliostat	Via heliostat	-
Solar irradiance	980 W/m^2	870 W/m^2	860 W/m^2	850 W/m^2	-

Table 1. *Cont.*

Parameters	Guan et al. 2018 [25]	Liang et al. 2018 [19]	Vistas et al. 2021 [22]	This Work 2022	Improvements Over Previous Record (Times)
Active medium Pumping method	Nd:YAG/YAG End-side-pump	Cr:Nd:YAG End-side-pump	Ce:Nd:YAG Side-pump	Ce:Nd:YAG End-side-pump	-
Laser power	31.1 W	32.5 W	16.5 W	11.2 W	-
Minimum incoming threshold power	**200 W**	400 W	220 W	88 W	**0.44** [25]
Solar-to-laser conversion efficiency	3.1%	**3.7%**	2.8%	4.5%	**1.22** [19]
Solar laser collection efficiency	32.1 W/m²	**32.5 W/m²**	23.6 W/m²	38.22 W/m²	**1.18** [19]
Slope efficiency	5.4%	**6.7%**	4.4%	6.8%	**1.02** [19]

The laser beam M^2 quality factors of the Ce:Nd:YAG were measured according to ISO 11164-1 standards, with a CINOGY UV-NIR beam profiler using a CinCam CMOS. Figure 13a shows the multimode laser beam profile. The measured solar laser beam widths along the beam caustic and the associated extrapolated hyperbolic plot are shown in Figure 13b. $M_x^2 = 32.16 \pm 2.62$ and $M_y^2 = 36.62 \pm 4.22$ factors were experimentally obtained.

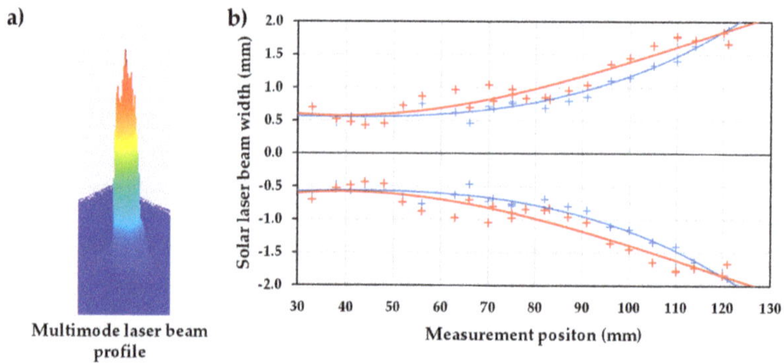

Figure 13. (**a**) Multimode beam profile. (**b**) Caustic fit measurements of the multimode solar laser beam along the X-axis (red color) and Y-axis (blue color).

6. Conclusions

The highly efficient Ce:Nd:YAG solar laser was composed of the first-stage heliostat-parabolic mirror solar energy collection and concentration system; the second-stage fused silica aspheric lens that further concentrated the solar energy onto the end-face of the active medium by end-pumping; and the third-stage conical-shaped reflective pump cavity to efficiently side-pump the 2.5 mm diameter, 25 mm length Ce:Nd:YAG rod. The solar energy transfer for the Nd^{3+} and Ce^{3+} ions as well as the energy transfer of Ce^{3+} to Nd^{3+} were introduced and considered in Zemax©. The design parameters of the laser head were optimized to find the highest solar energy absorption by the active medium to then determine the best lasing conditions in LASCAD™. The Ce:Nd:YAG solar laser prototype was then built according to the simulated model. 11.2 W multimode laser output power was successfully measured, matching well with the numerical result. This, to the best of our knowledge, resulted in a record solar-to-laser conversion efficiency of 4.50%. The highest solar laser collection and slope efficiencies of 38.22 W/m² and 6.8%, respectively, were also obtained. Furthermore, low threshold power of 88 W was also reported. Since the NOVA solar energy collection and concentration facility has a limited transfer efficiency of 0.75, further research on small size Ce:Nd:YAG solar laser in direct solar tracking mode will, most hopefully, ensure further solar laser efficiency enhancement and enable promising space applications.

Author Contributions: Conceptualization, D.G.; methodology, D.G.; software, D.G., D.L., C.R.V., H.C. and M.C.; validation, D.G. and D.L.; formal analysis, D.G., D.L., C.R.V. and J.A.; investigation, D.G.; resources, D.L. and J.A.; data curation, D.G. and D.L.; writing—original draft preparation, D.G.; writing—review and editing, D.G., D.L., C.R.V., M.C., B.D.T., H.C. and J.A.; visualization, D.G.; supervision, D.L.; project administration, D.L.; funding acquisition, D.L. All authors have read and agreed to the published version of the manuscript.

Funding: This research was funded by Fundação para a Ciência e a Tecnologia—Ministério da Ciência, Tecnologia e Ensino Superior, grant number UIDB/00068/2020.

Institutional Review Board Statement: Not applicable.

Informed Consent Statement: Not applicable.

Data Availability Statement: Not applicable.

Acknowledgments: The contract CEECIND/03081/2017 and the fellowship grants SFRH/BD/145322/2019, PD/BD/142827/2018, PD/BD/128267/2016 and SFRH/BPD/125116/2016 of Joana Almeida, Miguel Catela, Dário Garcia, Bruno D. Tibúrcio, and Cláudia R. Vistas, respectively, are acknowledged.

Conflicts of Interest: The authors declare no conflict of interest. The funders had no role in the design of the study; in the collection, analyses, or interpretation of data; in the writing of the manuscript, or in the decision to publish the results.

Nomenclature

A	Collection area (m^2)
c	Radius of curvature (mm)
I	Solar irradiance (W/m^2)
k	Parabolic constant
$P_{abs,ion}$	Absorbed solar power by ion (W)
L_c	Length rod (mm)
$L_{AR\text{-}PR}$	Distance between AR and PR (mm)
r	Radial aperture (mm)
R	Reflectivity
z	Sag (mm)
Greek symbol	
α_{ion}	Absorption coefficient (cm^{-1})
β_1	Aspheric coefficient (mm^{-1})
$\eta_{overlap,ion}$	Overlap efficiency
$\eta_{NR:Ce\text{-}Nd}$	Non-radiative conversion efficiency of Ce^{3+} to Nd^{3+}
λ	Wavelength (nm)
Abbreviation	
AR	Anti-reflection
HR	High reflection
LASCAD™	Laser cavity analysis and design
MSSF	Medium sized solar furnace
PR	Partial reflection
RoC	Radius of curvature
YAG	Yttrium aluminum garnet

References

1. The Global Goals. Goal 7: Affordable and Clean Energy. Available online: https://www.globalgoals.org/goals/7-affordable-and-clean-energy/ (accessed on 1 April 2022).
2. The Global Goals. Goal 9: Industry, Innovation and Infrastructure. Available online: https://www.globalgoals.org/goals/9-industry-innovation-and-infrastructure/ (accessed on 1 April 2022).
3. Motohiro, T.; Takeda, Y.; Ito, H.; Hasegawa, K.; Ikesue, A.; Ichikawa, T.; Higuchi, K.; Ichiki, A.; Mizuno, S.; Ito, T.; et al. Concept of the solar-pumped laser-photovoltaics combined system and its application to laser beam power feeding to electric vehicles. *Jpn. J. Appl. Phys.* **2017**, *56*, 08MA07. [CrossRef]
4. Yabe, T.; Bagheri, B.; Ohkubo, T.; Uchida, S.; Yoshida, K.; Funatsu, T.; Oishi, T.; Daito, K.; Ishioka, M.; Yasunaga, N.; et al. 100 W-class solar pumped laser for sustainable magnesium-hydrogen energy cycle. *J. Appl. Phys.* **2008**, *104*, 083104. [CrossRef]
5. Abdel-Hadi, Y.A. Space-based solar laser system simulation to transfer power onto the earth. *NRIAG J. Astron. Geophys.* **2020**, *9*, 558–562. [CrossRef]
6. Kiss, Z.J.; Lewis, H.R.; Duncan, R.C. Sun pumped continuous optical maser. *Appl. Phys. Lett.* **1963**, *2*, 93–94. [CrossRef]
7. Insuik, R.J.; Christiansen, W.H. Blackbody-pumped CO_2 laser experiment. *AIAA J.* **1984**, *22*, 1271–1274. [CrossRef]
8. Terry, C.K.; Peterson, J.E.; Goswami, D.Y. Terrestrial solar-pumped iodine gas laser with minimum threshold concentration requirements. *J. Thermophys. Heat Transf.* **1996**, *10*, 54–59. [CrossRef]
9. Lee, J.H.; Kim, K.C.; Kim, K.H. Threshold pump power of a solar-pumped dye laser. *Appl. Phys. Lett.* **1988**, *53*, 2021–2022. [CrossRef]
10. Lando, M.; Kagan, J.; Linyekin, B.; Dobrusin, V. A solar-pumped Nd: YAG laser in the high collection efficiency regime. *Opt. Commun.* **2003**, *222*, 371–381. [CrossRef]
11. Weksler, M.; Shwartz, J. Solar-Pumped Solid-State Lasers. *IEEE J. Quantum Electron.* **1988**, *24*, 1222–1228. [CrossRef]
12. Yabe, T.; Ohkubo, T.; Uchida, S.; Yoshida, K.; Nakatsuka, M.; Funatsu, T.; Mabuti, A.; Oyama, A.; Nakagawa, K.; Oishi, T.; et al. High-efficiency and economical solar-energy-pumped laser with Fresnel lens and chromium codoped laser medium. *Appl. Phys. Lett.* **2007**, *90*, 261120. [CrossRef]
13. Liang, D.; Almeida, J. Highly efficient solar-pumped Nd:YAG laser. *Opt. Express* **2011**, *19*, 26399–26405. [CrossRef] [PubMed]
14. Dinh, T.H.; Ohkubo, T.; Yabe, T.; Kuboyama, H. 120 watt continuous wave solar-pumped laser with a liquid light-guide lens and an Nd:YAG rod. *Opt. Lett.* **2012**, *37*, 2670–2672. [CrossRef] [PubMed]
15. Zhao, B.; Zhao, C.; He, J.; Yang, S. Study of active medium for solar-pumped solid-state lasers. *Guangxue Xuebao Acta Opt. Sin.* **2007**, *27*, 1797–1801.
16. Yagi, H.; Yanagitani, T.; Yoshida, H.; Nakatsuka, M.; Ueda, K.-I. Highly Efficient Flashlamp-Pumped Cr^{3+} and Nd^{3+} Codoped $Y_3Al_5O_{12}$ Ceramic Laser. *Jpn. J. Appl. Phys.* **2006**, *45*, 133–135. [CrossRef]
17. Payziyev, S.; Makhmudov, K.; Abdel-Hadi, Y.A. Simulation of a new solar Ce:Nd:YAG laser system. *Optik* **2018**, *156*, 891–895. [CrossRef]
18. Liang, D.; Almeida, J.; Guillot, E. Side-pumped continuous-wave Cr:Nd:YAG ceramic solar laser. *Appl. Phys. B* **2013**, *111*, 305–311. [CrossRef]
19. Liang, D.; Vistas, C.R.; Tibúrcio, B.D.; Almeida, J. Solar-pumped Cr:Nd:YAG ceramic laser with 6.7% slope efficiency. *Sol. Energy Mater. Sol. Cells* **2018**, *185*, 75–79. [CrossRef]
20. Vistas, C.R.; Liang, D.; Garcia, D.; Almeida, J.; Tibúrcio, B.D.; Guillot, E. Ce:Nd:YAG continuous-wave solar-pumped laser. *Optik* **2020**, *207*, 163795. [CrossRef]
21. Li, Y.; Zhou, S.; Lin, H.; Hou, X.; Li, W. Intense 1064 nm emission by the efficient energy transfer from Ce^{3+} to Nd^{3+} in Ce/Nd co-doped YAG transparent ceramics. *Opt. Mater.* **2010**, *32*, 1223–1226. [CrossRef]
22. Vistas, C.R.; Liang, D.; Almeida, J.; Tibúrcio, B.D.; Garcia, D.; Catela, M.; Costa, H.; Guillot, E. Ce:Nd:YAG side-pumped solar laser. *J. Photonics Energy* **2021**, *11*, 1–9. [CrossRef]
23. Vistas, C.R.; Liang, D.; Garcia, D.; Catela, M.; Tibúrcio, B.D.; Costa, H.; Guillot, E.; Almeida, J. Uniform and Non-Uniform Pumping Effect on Ce:Nd:YAG Side-Pumped Solar Laser Output Performance. *Energies* **2022**, *15*, 3577. [CrossRef]
24. Almeida, J.; Liang, D.; Garcia, D.; Tibúrcio, B.D.; Costa, H.; Catela, M.; Guillot, E.; Vistas, C.R. 40 W Continuous Wave Ce:Nd:YAG Solar Laser through a Fused Silica Light Guide. *Energies* **2022**, *15*, 3998. [CrossRef]
25. Guan, Z.; Zhao, C.; Li, J.; He, D.; Zhang, H. 32.1 W/m^2 continuous wave solar-pumped laser with a bonding Nd:YAG/YAG rod and a Fresnel lens. *Opt. Laser Technol.* **2018**, *107*, 158–161. [CrossRef]
26. Holloway, W.W.; Kestigian, M. Optical Properties of Cerium-Activated Garnet Crystals. *J. Opt. Soc. Am.* **1969**, *59*, 60–63. [CrossRef]
27. Mares, J.; Jacquier, B.; Pédrini, C.; Boulon, G. Energy transfer mechanisms between Ce^{3+} and Nd^{3+} in YAG: Nd, Ce at low temperature. *Rev. Phys. Appl.* **1987**, *22*, 145–152. [CrossRef]
28. Tai, Y.; Zheng, G.; Wang, H.; Bai, J. Near-infrared quantum cutting of Ce^{3+}–Nd^{3+} co-doped $Y_3Al_5O_{12}$ crystal for crystalline silicon solar cells. *J. Photochem. Photobiol. A Chem.* **2015**, *303–304*, 80–85. [CrossRef]
29. Powell, R.C. *Physics of Solid-State Laser Materials*; Springer: New York, NY, USA, 1998.
30. *ASTM G173-03(2012)*; Standard Tables for Reference Solar Spectral Irradiances: Direct Normal and Hemispherical on 37° Tilted Surface. ASTM International: West Conshohocken, PA, USA, 2012.

31. Payziyev, S.; Sherniyozov, A.; Bakhramov, S.; Zikrillayev, K.; Khalikov, G.; Makhmudov, K.; Ismailov, M.; Payziyeva, D. Luminescence sensitization properties of Ce: Nd: YAG materials for solar pumped lasers. *Opt. Commun.* **2021**, *499*, 127283. [CrossRef]
32. Ueda, J.; Tanabe, S. (INVITED) Review of luminescent properties of Ce3+-doped garnet phosphors: New insight into the effect of crystal and electronic structure. *Opt. Mater. X* **2019**, *1*, 100018. [CrossRef]
33. Samuel, P.; Yanagitani, T.; Yagi, H.; Nakao, H.; Ueda, K.I.; Babu, S.M. Efficient energy transfer between Ce^{3+} and Nd^{3+} in cerium codoped Nd: YAG laser quality transparent ceramics. *J. Alloys Compd.* **2010**, *507*, 475–478. [CrossRef]
34. Yamaga, M.; Oda, Y.; Uno, H.; Hasegawa, K.; Ito, H.; Mizuno, S. Energy transfer from Ce to Nd in $Y_3Al_5O_{12}$ ceramics. *Phys. Status Solidi C* **2012**, *9*, 2300–2303. [CrossRef]
35. LAS-CAD GmbH. *LASCAD, 3.3.5 Manual*; LAS-CAD GmbH: Munchen, Germany, 2007.
36. Dong, J.; Rapaport, A.; Bass, M.; Szipocs, F.; Ueda, K.-I. Temperature-dependent stimulated emission cross section and concentration quenching in highly doped Nd^{3+}: YAG crystals. *Phys. Status Solidi A* **2005**, *202*, 2565–2573. [CrossRef]

Article

Doughnut-Shaped and Top Hat Solar Laser Beams Numerical Analysis

Miguel Catela, Dawei Liang *, Cláudia R. Vistas, Dário Garcia, Bruno D. Tibúrcio, Hugo Costa and Joana Almeida

Center for Physics and Technological Research (CEFITEC), Department of Physics, Faculty of Science and Technology, NOVA University of Lisbon, 2829-516 Caparica, Portugal; m.catela@campus.fct.unl.pt (M.C.); c.vistas@fct.unl.pt (C.R.V.); kongming.dario@gmail.com (D.G.); brunotiburcio78@gmail.com (B.D.T.); hf.costa@campus.fct.unl.pt (H.C.); jla@fct.unl.pt (J.A.)
* Correspondence: dl@fct.unl.pt

Abstract: Aside from the industry-standard Gaussian intensity profile, top hat and non-conventional laser beam shapes, such as doughnut-shaped profile, are ever more required. The top hat laser beam profile is well-known for uniformly irradiating the target material, significantly reducing the heat-affected zones, typical of Gaussian laser irradiation, whereas the doughnut-shaped laser beam has attracted much interest for its use in trapping particles at the nanoscale and improving mechanical performance during laser-based 3D metal printing. Solar-pumped lasers can be a cost-effective and more sustainable alternative to accomplish these useful laser beam distributions. The sunlight was collected and concentrated by six primary Fresnel lenses, six folding mirror collectors, further compressed with six secondary fused silica concentrators, and symmetrically distributed by six twisted light guides around a 5.5 mm diameter, 35 mm length Nd:YAG rod inside a cylindrical cavity. A top hat laser beam profile ($M_x^2 = 1.25$, $M_y^2 = 1.00$) was computed through both ZEMAX® and LASCAD® analysis, with 9.4 W/m² TEM$_{00}$ mode laser power collection and 0.99% solar-to-TEM$_{00}$ mode power conversion efficiencies. By using a 5.8 mm laser rod diameter, a doughnut-shaped solar laser beam profile ($M_x^2 = 1.90$, $M_y^2 = 1.00$) was observed. The 9.8 W/m² TEM$_{00}$ mode laser power collection and 1.03% solar-to-TEM$_{00}$ mode power conversion efficiencies were also attained, corresponding to an increase of 2.2 and 1.9 times, respectively, compared to the state-of-the-art experimental records. As far as we know, the first numerical simulation of doughnut-shaped and top hat solar laser beam profiles is reported here, significantly contributing to the understanding of the formation of such beam profiles.

Keywords: solar laser; solar pumping; twisted light guide; Nd:YAG; top hat; doughnut-shaped

Citation: Catela, M.; Liang, D.; Vistas, C.R.; Garcia, D.; Tibúrcio, B.D.; Costa, H.; Almeida, J. Doughnut-Shaped and Top Hat Solar Laser Beams Numerical Analysis. *Energies* **2021**, *14*, 7102. https://doi.org/10.3390/en14217102

Academic Editor: Tapas Mallick

Received: 9 September 2021
Accepted: 28 October 2021
Published: 31 October 2021

1. Introduction

The direct conversion of sunlight into narrow-band coherent laser radiation represents a cost-effective innovation in laser technology via renewable energy. Compared to indirect solar-to-laser conversion by using solar panels, direct solar-to-laser conversion eliminates the intermediate conversion process from solar to electricity and thus may be inherently more efficient, much simpler, and more reliable [1]. Since most of the electronics are no longer required in direct solar-to-laser conversion, potentially limiting problems, for example, related to high voltages [2], can be eliminated. Solar lasers can hence be intently more cost-effective than classical electricity-powered lasers. The dismissal of electrical pumping sources, such as arc lamps and laser diodes, along with their mandatory power consumption and conditioning equipment, constitutes an added value. This makes solar lasers potentially well suited for space-based applications, including deep-space optical communications [3], solar power transmission [4], laser propulsion [5], and asteroid deflection [6]. Additionally, they can be a valuable asset for industries that depend on materials processing at high temperatures [7–9].

The efficient extraction of TEM_{00} mode solar laser power is of utmost importance for many laser-based applications. Its Gaussian laser beam profile can be focused into a diffraction-limited spot with the highest energy density and the lowest beam divergence [10]. Therefore, it is widely applied in the materials processing industry [11]. The irradiance cross-section of Gaussian beams, however, gradually decreases from the center to the periphery of the laser spot so that a portion of the laser beam profile may not have enough irradiance for the given application. Moreover, this wasted energy can even damage surrounding areas outside of the target, extending the heat-affected zones [12]. The simplest way to address this issue is to convert the Gaussian energy distribution into a more uniform profile, such as the top hat beam profile [13]. For this reason, several beam shaping techniques are now on the market: reflective [14] and refractive [15–17] configurations, diffractive interference-based models [18–20], that can be implemented through digital micromirrors [21] or spatial light modulators [22], intra-cavity modulation [23,24], beam integrators [25] by using Powell [26] or even freeform lenses [27], and via optical fibers with a modified core [28].

Ideally, a top hat beam profile has zero energy at its edges and a constant energy distribution through its cross-section. Experimentally, it is not possible to obtain such an idyllic top hat beam since it would require an infinite spatial frequency spectrum [13], but several approximations can be made, namely Fermi-Dirac, super-Lorentzian, super-Gaussian, flattened Gaussian beams, and multi-Gaussian beams [29,30]. Compared to Gaussian beams, a more flattened beam profile can generate cleaner cuts and sharper edges, resulting in increased accuracy for high-demand applications [31], including laser micro-machining [12], precise materials processing [32], direct laser interference patterning [33] and precise laser surgery [34]. However, generating a top hat profile raises the system cost and complexity while its output power is significantly reduced. Moreover, top hat beam profiles do not preserve their flattened intensity during propagation through an optical system, wilting into the well-known "Airy disk" distribution [13].

Aside from the typical Gaussian intensity profile and the highly accurate top hat profile, non-conventional laser beam shapes are ever more required for specific applications [35]. Annular beams, in particular the doughnut-shaped profile (TEM_{01}^{*} mode), may broaden the laser technology applications, especially at nanoscale, extending laser materials processing techniques, improving lithography accuracy, and creating novel structures in materials [11,36]. Furthermore, their annular thermal profile is significantly important when the temperature is the leading factor, specifically in laser heat treatment and laser hardening. Here, solar-pumped lasers can be a cost-effective and more sustainable alternative.

Still, the pump power necessary to initiate laser emission requires a collection and concentration system. Parabolic mirrors have been used since the first solar-pumped Nd:YAG laser developed by Young in 1966 [37], shortly after the laser invention itself [38]. The high flux achieved with parabolic mirror has guided other solar laser researchers to explore it as a primary concentrator [39–43]. In 2017, 31.5 W/m^2 solar laser collection efficiency, described as the solar laser output power per unit of primary concentrator area [41], was attained by pumping an Nd:YAG laser rod with 4 mm in diameter and 35 mm in length through a parabolic mirror with an effective collection area of 1.18 m^2 [42]. Moreover, record fundamental mode solar laser collection efficiency of 7.9 W/m^2 was attained at the solar facility of the Procédés, Matériaux et Énergie Solaire—Centre National de la Recherche Scientifique [42].

Even though parabolic mirrors can reach high solar fluxes, their heavy weight and high cost stand as a barrier for experimental solar laser research and future applications. Furthermore, their overall efficiency is significantly reduced because the receiver and the corresponding mechanical fixation support have to be placed between the incoming solar radiation and the solar concentrator, creating shadows. Contrarily, the Fresnel lenses focusing method does not originate shadows. In addition, their low weight, affordable cost, availability in large size, and adequacy for mass production make them a popular, cost-effective primary concentrator for solar laser research [1,44–48]. The 2.93 W/m^2 solar laser collection efficiency in the TEM_{00} mode regime and 0.33% solar-to-TEM_{00} mode power conversion efficiency were attained in 2013 by pumping a 3 mm diameter, 30 mm length Nd:YAG laser rod with a 0.785 m^2 Fresnel lens [46]. In 2018, the solar laser collection efficiency of 32.1 W/m^2 was experimentally achieved by pumping a 6 mm diameter, 95 mm length Nd:YAG/YAG composite rod through a Fresnel lens with a 1.03 m^2 collection area [48]. More recently, a novel seven-rod/seven-beam pumping concept by a 4.0 m^2 Fresnel lens was modeled, with 13.66 W/m^2 TEM_{00} mode solar laser collection and 1.44% solar-to-TEM_{00} mode power conversion efficiencies being numerically obtained [49]. End-side pumping configurations have reached records in solar laser efficiency [1,42–45,47–49]. Still, the side-pumping approach may lead to higher laser beam brightness, defined as the ratio between the solar laser output power and the product of the beam quality factors M_x^2 and M_y^2 [41], as the solar pump radiation can be uniformly absorbed along the rod axis, originating a smoother temperature profile and preventing thermal induced effects. Additionally, both rod ends can be more easily accessed, enabling the optimization of more resonant cavity parameters and consequently leading to improvements in the laser beam quality and a more efficient TEM_{00} mode laser beam extraction. With the side-pumping configuration, the current records for multimode solar laser collection and solar-to-multimode power conversion efficiencies are 17.0 W/m^2 and 2.43%, respectively [50], whereas, for TEM_{00} mode laser power, experimental records of 3.1 W/m^2 collections and 0.40% conversion efficiencies were achieved [50]. The 3.2 W/m^2 TEM_{01}^* mode solar-to-laser collection efficiency and 0.46% solar-to-TEM_{01}^* mode power conversion efficiency were also obtained [50]. The production of this non-conventional laser beam with direct solar conversion showed the remarkable versatility that solar laser systems can offer. The doughnut-shaped solar laser beam was first achieved experimentally by Almeida et al. in 2018. The 2.7 W/m^2 collection efficiency was reported by side-pumping a grooved Nd:YAG rod with the NOVA heliostat-parabolic mirror system [51]. Vistas et al. attained 4.5 W/m^2 record collection efficiency for doughnut-shaped solar laser beam by end-pumping an Nd:YAG laser rod within a conical cavity [52]. The selective oscillation of certain laser modes is influenced by the thermal lens effects on the active medium and spatial mode-matching efficiency [51]. This is a simpler and cheaper approach with no need for additional optical elements, such as apertures, intra-cavity phase components, or spatial light modulators [53–55], that significantly deplete the output laser power. Table 1 presents a summary of the above-described solar-pumped laser beam achievements in collection and conversion efficiencies, as well as in and laser beam quality factors.

Table 1. Summary of solar-pumped laser beam achievements in collection and conversion efficiencies and beam quality factors.

Scheme	Primary Concentrator	Laser Mode	Collection Efficiency (W/m^2)	Conversion Efficiency (%)	Beam Quality Factors
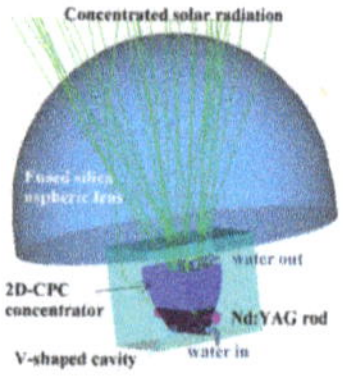 Liang et al., 2013 [46] (Experimental)	Fresnel lens	Multimode	10.3	1.16	Not reported
		TEM$_{00}$	2.93	0.64	$M_x^2 \approx M_y^2 < 1.1$
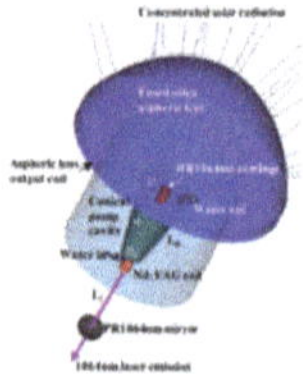 Liang et al., 2017 [42] (Experimental)	Parabolic mirror	Multimode	31.5	2.4	$M_x^2 \approx M_y^2 = 53.4$
		TEM$_{00}$	7.9	0.60	$M_x^2 \approx M_y^2 < 1.2$
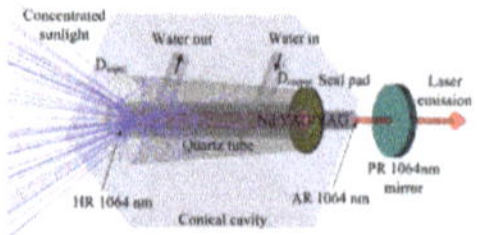 Guan et al., 2018 [48] (Experimental)	Fresnel lens	Multimode	32.1	3.31	$M_x^2 \approx M_y^2 = 61.0$
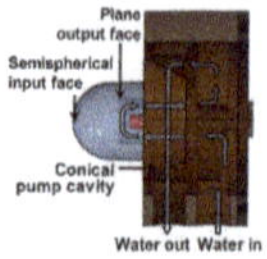 Vistas et al., 2018 [52] (Experimental)	Parabolic mirror	TEM$_{01}{}^*$	4.5	0.54	Not reported
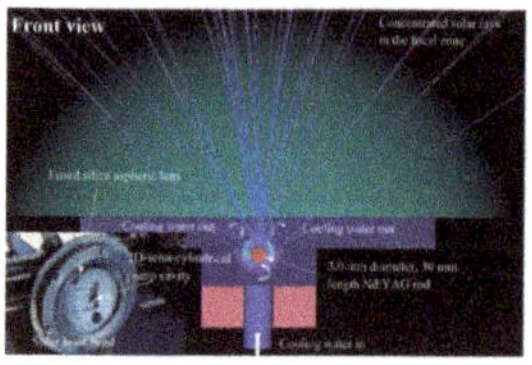 Liang et al., 2019 [50] (Experimental)	Parabolic mirror	Multimode	17.0	2.43	$M_x^2 \approx M_y^2 = 16.8$
		TEM$_{00}$	3.1	0.40	Not reported
		TEM$_{01}{}^*$	3.2	0.46	Not reported
 Liang et al., 2021 [49] (Numerical)	Fresnel lens	Multimode	23.3	2.82	Not reported
		TEM$_{00}$	13.7	1.44	$M_x^2 = 1.00$ $M_y^2 = 1.04$

In the present solar laser system, the solar energy was both collected and concentrated by six Fresnel lenses, with a 4.0 m^2 total collection area, and redirected into the laser head through six plane folding mirrors. Six secondary fused silica concentrators further compressed the solar radiation onto the square input face of six twisted fused silica light guides. These light guides were essential to transform the near-Gaussian energy distribution from each aspheric concentrator into a narrow rectangular pump column. Hence, it was possible to closely couple the solar radiation symmetrically around a 5.5 mm diameter, 35 mm length Nd:YAG rod within a cylindrical cavity with water cooling. The laser resonator was composed of a 1064 nm high reflection (HR)-coated end mirror and a 1064-nm partial reflection (PR)-coated output coupler. Its asymmetrical configuration ensured the best overlap between the solar pump radiation and the laser mode volumes. This novel solar laser system was able to numerically simulate top hat and doughnut-shaped laser beams, which are of major importance for applications that simultaneously require accuracy and high-temperature processing while competing in efficiency with the most advanced solar laser systems with side-pumping configuration [50]. As far as we know, no numerical simulation of doughnut-shaped and top hat solar laser beams was previously reported. Even though doughnut-shaped solar laser beams were experimentally obtained, their production methods were not fully exploited. Hence, the first numerical simulation of doughnut-shaped and top hat solar laser beam profiles reported here significantly contributes to understanding how to generate such beams, particularly the top hat solar laser beam that was not yet demonstrated in practical essays. In addition, we report an increase of 2.2 and 1.9 times in collection and solar-to-laser conversion efficiencies, respectively, compared to the state-of-the-art experimental records for doughnut-shaped solar laser beams [52].

2. Methods

2.1. Solar Energy Collection and Concentration System with Six Fresnel Lenses

The proposed side-pumping solar laser approach (Figure 1) was composed of six circular Fresnel lenses (F_1–F_6) for collection and concentration of the incoming solar radiation, aligned with six plane folding mirrors (M_1–M_6), which redirected the concentrated solar radiation from the Fresnel lenses onto the laser head. Each Fresnel lens was centered at a distance d = 950 mm from their common optical center point C, which was placed at a height h = 551 mm above the center of the laser head, as indicated in Figure 1. The plane folding mirrors had an inclination angle of 45° in relation to their common optical axis and were located below their respective Fresnel lenses. In ZEMAX® non-sequential analysis, the radius parameter defines the radius of curvature of the Fresnel lens. For this system, an optimum radius of curvature of 700 mm was used, corresponding to a focal length of approximately 1.5 m. The Fresnel lenses were made of polymethyl methacrylate, which has high transmission efficiency for both visible and near-infrared wavelengths between 350 and 900 nm. Each one of the 3 mm thick Fresnel lenses had a radius of 461 mm, summing a total collection area of 4 m^2. For the whole solar spectrum, an averaged transmission efficiency of 84% was numerically attained for the Fresnel lenses. For 950 W/m^2 terrestrial solar irradiance and 95% reflectivity plane folding mirrors, a total of 3032 W solar power was assumed to reach the laser head.

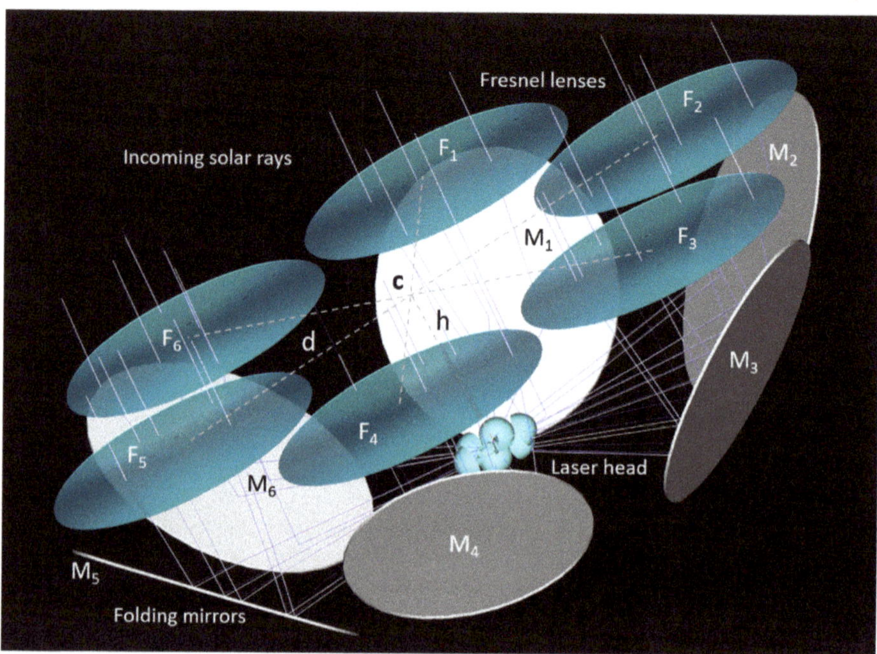

Figure 1. Six Fresnel lens solar laser side-pumping concept with 4.0 m^2 total collection area. F_1–F_6 and M_1–M_6 indicate the six Fresnel lenses and the six folding mirrors, respectively. Each Fresnel lens was symmetrically positioned at a distance d = 950 mm from their common optical center point C and at a height of h = 551 mm from the center of the laser head.

2.2. Solar Laser Head with Six Aspheric Lenses and Six Twisted Light Guides

For a high focusing of the solar radiation into the Nd:YAG laser rod, six fused silica aspheric concentrators, and six twisted fused silica light guides were designed, as presented in Figure 2. Fused silica with high quality (99.999%) can be obtained through optical machining and polishing [56]. Its transparency over the visible region of the solar emission spectrum, low coefficient of thermal expansion, and high resistance to scratching and thermal endurance are optimal characteristics for Nd:YAG laser rod pumping. The six aspheric lenses were all equally designed to efficiently couple the concentrated solar radiation from the focal zone of each Fresnel lens onto the input face of each twisted light guide. Each optimized aspheric concentrator had 112 mm diameter, 53 mm height, −53 mm rear radius, and −0.150 rear conic value, and all were symmetrically positioned 96.5 mm away from the center of the laser head. An averaged transmission efficiency of 87% was numerically determined for the fused silica aspheric lenses, each one focusing 440 W solar power.

The concentrated solar radiation was then transmitted through six twisted fused silica light guides with 10.00 × 10.00 mm input face, 62.50 mm length, and 3.33 × 30.00 mm output end, experiencing the principles of refraction and total internal reflection, as demonstrated in Figures 3 and 4. The six twisted light guides were designed with AutoCAD software and evenly redistributed at 4 mm from the center of the laser head in order to deliver efficiently the concentrated solar radiation from each aspheric lens. Figure 3 shows that each one of the twisted light guides was composed of five sections: the 3.33 × 10.00 mm central section, two 3.33 × 6.67 mm sections, and two 3.33 × 3.33 mm sections. As illustrated in Figure 3 inset, at the guide's input face, the five sections were arranged in a 10.0 × 10.00 mm square cross-section, and at its output end, they became a single column with 3.33 × 30.00 mm rectangular cross-section.

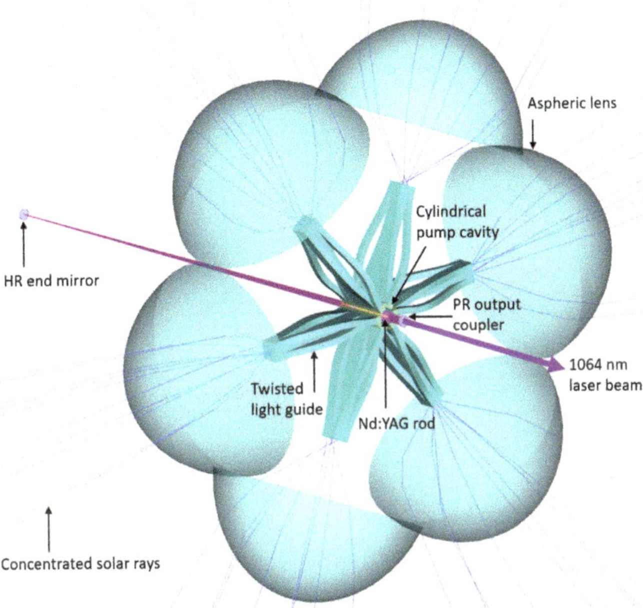

Figure 2. Three-dimensional view of the six secondary aspheric lenses, the six twisted light guides, the cylindrical pump cavity, and the Nd:YAG rod, and the high reflection (HR) end mirror and the partial reflection (PR) output coupler, which compose the resonant cavity.

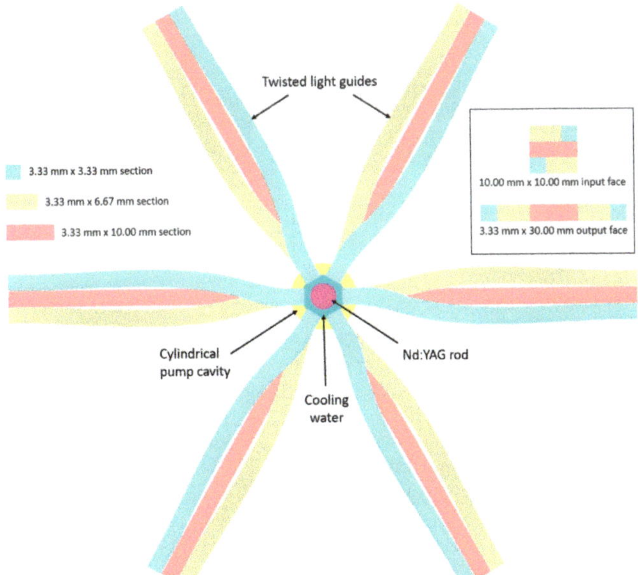

Figure 3. Cross-sectional view of the six twisted light guides assembled for distributing the concentrated solar radiation from the aspheric lenses into the laser rod. Each one of the twisted light guides was composed of five sections: the 3.33×10 mm central section, two 3.33×6.67 mm sections, and two 3.33×3.33 mm sections. The inset represents the top-view of the light guide input and output faces.

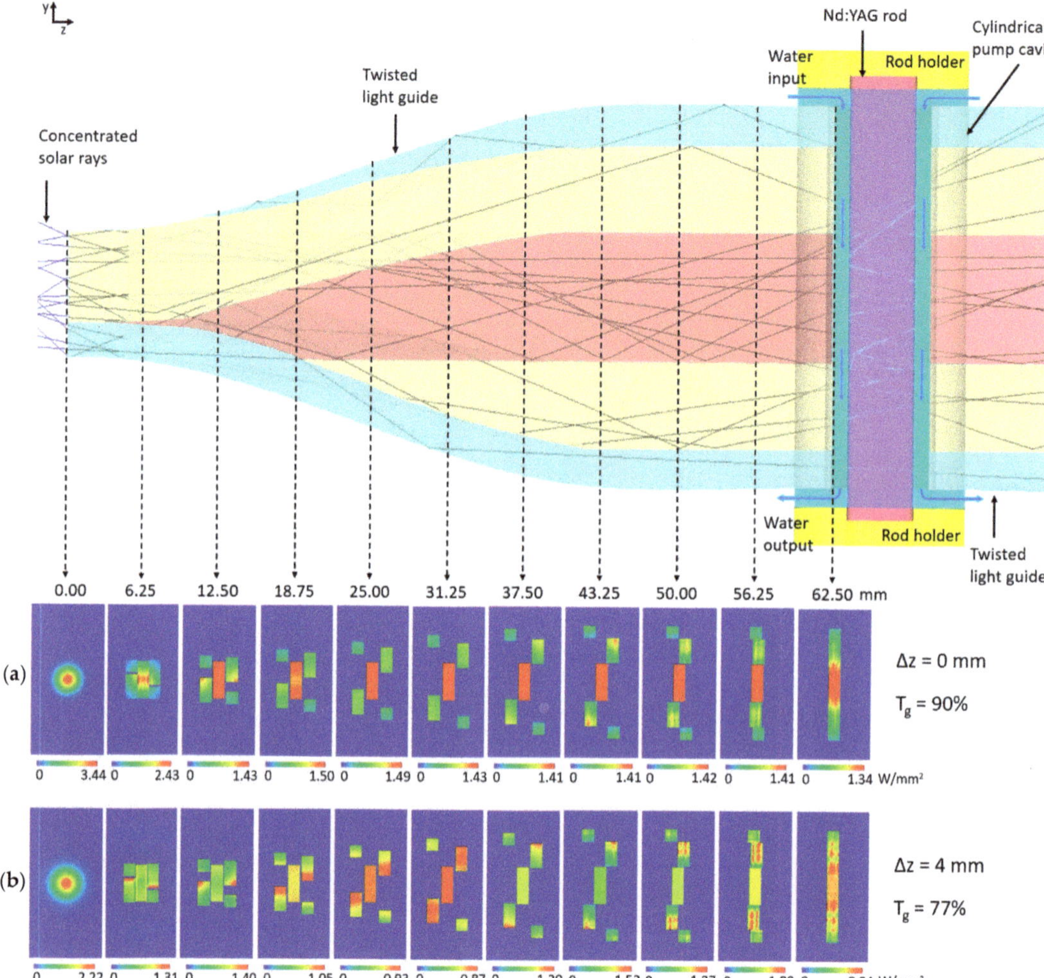

Figure 4. Pump light transmission through the twisted light guides. Only one twisted light guide is entirely shown to indicate how the solar rays were delivered into the laser rod inside the cylindrical pump cavity with cooling water. Pump light distribution at several cross-sections along one twisted light guide is also presented for the case of (**a**) optimum alignment between each aspheric lens and the input face of each light guide, and (**b**) shifting of the aspheric lens 4 mm upwards, resulting in a reduced transmission efficiency (T_g). Notice that the figure is not at scale.

Figure 4 gives the pump light distribution at several cross-sections for one twisted light guide. Maximum transmission is indicated with red color, whereas blue means almost none or zero pump light transmission. In order to obtain the maximum energy transmission, the optimum alignment between each aspheric lens and the input face of each light guide in the z-axis ($\Delta z = 0$) was determined. In this case, an averaged transmission efficiency of 90% was computed for the twisted light guides, each one delivering 396 W solar power into the cylindrical cavity (396 W × 6 = 2376 W total solar power). The contribution of the water layer was also included in the simulations, with the laser rod receiving at its surface a total of 1995 W concentrated solar power. However, about 50% of the energy was transmitted through the central section of each light guide, producing a non-uniform output power distribution, as demonstrated in Figure 4a. By shifting each aspheric lens by 4 mm upwards, it was possible to find a more uniform power distribution at the light guide's output end (Figure 4b), but the misalignment significantly reduced the light guide's averaged transmission efficiency to only 77%, and thus the shifting was not adopted.

As shown in Figures 3 and 4, a single 5.5 mm diameter, 35.0 mm length Nd:YAG rod was placed at the center of a cylindrical pump cavity (95% reflectivity inner surface) with a 9 mm internal diameter and 30.0 mm height. The Nd:YAG rod was cooled with water flowing inside the pump cavity along 32.5 mm of its longitudinal surface and was mechanically fixed on each side by two holders. A total of 1064 nm antireflection (AR) coating was considered for both Nd:YAG rod end faces. The 1064 nm HR mirror and the PR output coupler formed the laser resonator, as indicated in Figure 2.

Different laser active materials can be used in order to directly convert broad-band sunlight into monochromatic radiation [57–59]. Despite the correlation between the solar spectrum and the absorption spectrum of Nd:YAG being only 16%, its extraordinary spectroscopical properties [60] and great thermomechanical characteristics [40] that provide resilience and durability, combined with its easy availability and relatively low cost, have attracted researchers to use this laser material in the extreme thermal and optical conditions of solar pumping. Moreover, the Nd:YAG laser is a four-level laser system. This means that the lower laser level is well above the ground state and is quickly depopulated by multiple phonon transitions, avoiding reabsorption of the laser radiation, reducing the threshold pump power, and consequently facilitating stimulated emission. As presented in Figure 5, the laser photons are typically emitted at a wavelength of 1064 nm, corresponding to the transition from the $^4F_{3/2}$ to the $^4I_{11/2}$ level [10].

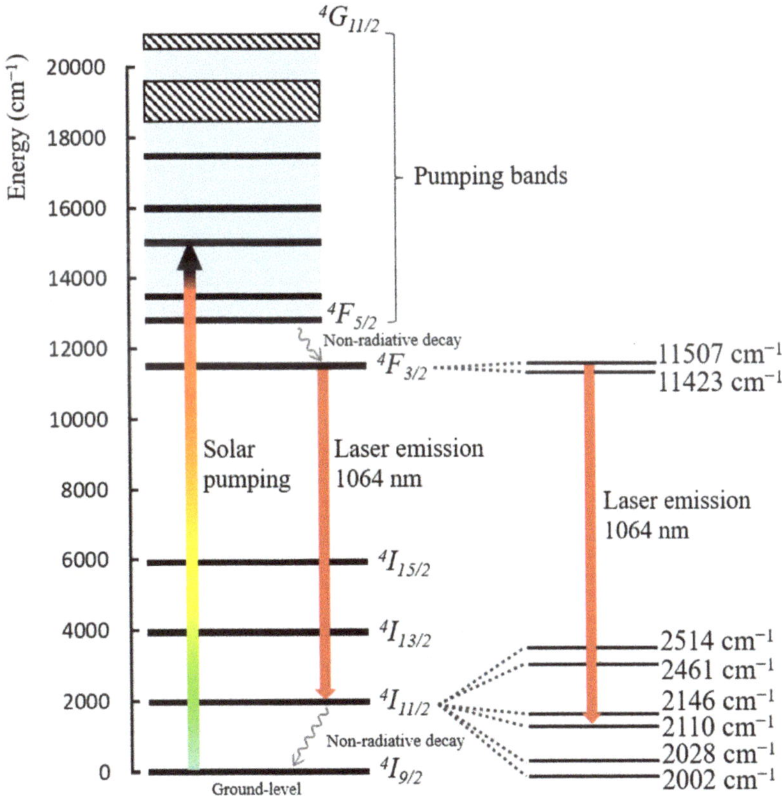

Figure 5. Simplified energy level diagram of Nd:YAG (Adapted from [10]).

2.3. Solar Laser Head Design with ZEMAX® Non-Sequential Ray-Tracing Software

Similar to previous research [42,43,46], the earlier-described design parameters of the side-pumping solar laser scheme were first optimized by the optical design software ZEMAX® software to attain maximal absorption of the pump power by the laser rod while trying to ensure a relatively smooth distribution of pump light. The 950 W/m^2 solar irradiance at Earth's surface, 16% overlap between Nd:YAG medium absorption and the solar emission spectra [61], 84% transmission efficiency through the Fresnel lenses, and 95% reflectance on the plane folding mirrors and the cylindrical pump cavity were considered in ZEMAX®. Additionally, 22 peak absorption wavelengths [43] characteristic of 1.0% Nd:YAG active material were programmed in ZEMAX® numerical data. The solar spectral irradiance (W/(m^2 nm)) at each wavelength was consulted from the standard direct solar spectrum for one-and-a-half air mass [62]. Those 22 wavelengths, as well as their absorption coefficients, along with the absorption spectrum and the refractive indices of fused silica material for the aspheric concentrators and twisted light guides, and cooling water, were programmed into the glass catalog of ZEMAX® software.

The side-pumping configuration enabled a symmetric light distribution around the laser rod with zig-zag reflections within the cylindrical cavity. The twisted fused silica light guides were fundamental in order to efficiently couple the concentrated solar radiation from the six aspheric lenses into the small-diameter cylindrical pump cavity by transforming the near-Gaussian light distribution from each aspheric concentrator into a narrow rectangular column of light at the output face of each light guide, as shown in Figures 3 and 4a. This way, efficient absorption of the pump light was achieved despite the chromatic aberration of the Fresnel lenses. The pump power distribution within the active medium was analyzed by ray-tracing ZEMAX® software. A rectangular detector volume, composed of 18,000 voxels, was placed in order to entirely cover the active media. This number of voxels and a high amount of analysis rays were used for the purpose of attaining more accurate results and better image resolution. However, these two parameters had to be carefully programmed since they can highly impact the overall simulation running time. The software computes the path length of the analysis rays across each voxel regarding the absorption coefficient of the 1.0 at% Nd:YAG medium at each peak wavelength. By summing up the contributions of all voxels, a total of 317 W pump power was numerically attained with the 5.5 mm diameter, 35 mm length Nd:YAG laser rod. Figure 6 presents the absorbed energy flux distributions for its central transversal and longitudinal cross-sections. Near maximum absorption is indicated in red color, whereas blue means little or zero absorption. The Nd:YAG rod showed a rotational symmetric absorbed pump flux profile. The data were then exported from ZEMAX® software and imported to LASCAD® software, as explained in Section 2.4.

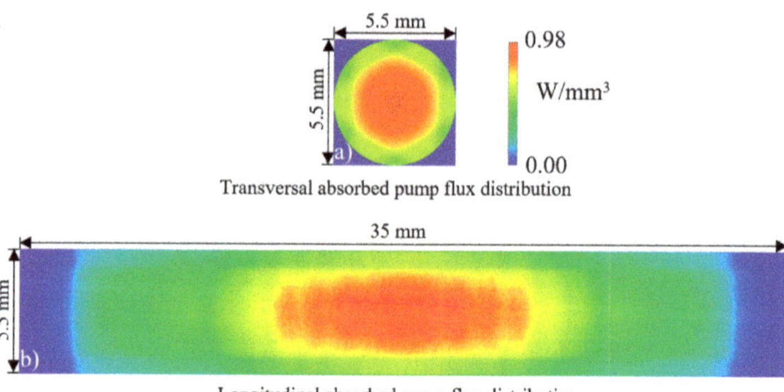

Figure 6. Absorbed pump flux distribution along both (**a**) transversal and (**b**) longitudinal cross-sections of the 5.5 mm diameter, 35 mm length Nd:YAG laser rod.

2.4. Laser Resonator Design through LASCAD® Software

The resonant cavity for the Nd:YAG laser rod was composed of two parallel mirrors: the HR end mirror, being 1064 nm HR-coated (99.98%), and the PR output coupler being 1064-nm PR-coated (between 80% and 99%, depending on the laser rod diameter for continuous-wave operation). A 1064-nm AR coating covered each rod end face. The intensity-weighted mean of the solar wavelengths over the laser absorption bands of 660 nm was adopted in LASCAD® analysis. For 1.0 at% Nd:YAG medium, the fluorescence lifetime of 230 µs, the stimulated emission cross-section of 2.8×10^{-19} cm^2, and the conventional absorption and scattering loss of 0.003 cm^{-1} were also considered [40].

The adoption of an asymmetric resonator significantly enhances the pump and laser mode matching, providing high-quality laser beams [63]. As shown in Figure 7, L_1 represents the distance from the HR mirror to the left end face of the laser rod, and L_2 the distance from the right end face of the laser rod to the PR output coupler, with both being particularly important parameters in order to obtain optimum mode overlap. L_2 was fixed at 10.0 mm, and L_1 was optimized by LASCAD® software in order to achieve the highest pump and laser mode matching. The laser rod length is represented by L_R. The radius of curvature of both the HR end mirror (RoC$_1$) and a PR output coupler (RoC$_2$) was also optimized. For the 5.5 mm diameter, 35 mm length rod, as indicated in Figure 7a, optimum laser design parameter was found at L_1 = 260.0 mm. At the output mirror, a top hat profile with a maximum power density of 6.0×10^{-6} W/µm^2 was verified, as shown in Figure 7b.

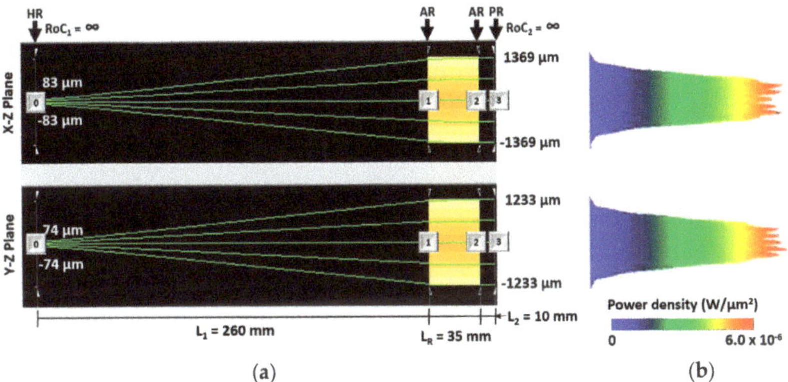

Figure 7. Laser cavity analysis with LASCAD® software: (**a**) Representation of the asymmetric laser cavity for efficient laser power extraction from the 5.5 mm diameter rod and (**b**) the respective beam profile. L_1 represents the distance from the HR mirror to the AR coating on the left rod end face, and L_2 is the distance from the AR coating on the right rod end face to the PR output coupler. The laser rod length is represented by L_R.

The beam waist was calculated by the beam propagation method (BPM) cavity iterations function of LASCAD® software for the top hat profile extracted from the 5.5 mm diameter laser rod within the laser cavity presented in Figure 7a. Figure 8 shows that the $1/e^2$ width beam waist value stabilizes at ω = 1233.0 µm.

In order to evaluate the laser beam behavior during propagation, the laser beam divergence half-angles through 10,000 mm vacuum were calculated at the X-Z and Y-Z planes, being θ_x = 0.011° and θ_y = 0.010°, respectively. Figure 9 presents the LASCAD® representation of the laser beam propagation. After 10,000 mm, the $1/e^2$ width beam waist value only increased from ω = 1233.0 µm to ω = 3013.0 µm.

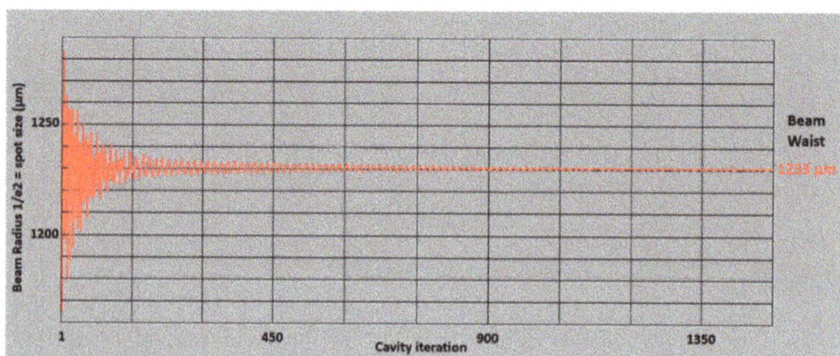

Figure 8. Laser beam waist at output mirror calculation by BPM cavity iterations function of LASCAD® software for the top hat profile extracted from the 5.5 mm diameter laser rod considering $L_1 = 260$ mm.

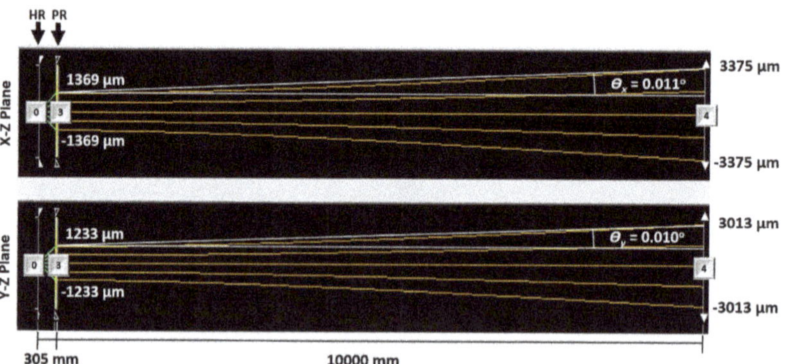

Figure 9. LASCAD® representation of the laser beam propagation through 10,000 mm vacuum. θ_x and θ_y represent the laser beam divergence half-angles at the X-Z and Y-Z planes, respectively.

3. Results

Based on the previously described parameters, multimode solar laser power, TEM_{00} mode laser power, and M^2 beam quality factors at the x and y-axis were calculated with the LASCAD® software. The numerical calculated TEM_{00} mode laser power corresponded to the fundamental mode within the obtained beam profile. The above-described laser resonant cavity was optimized for several laser rod diameters. Both the HR end mirror and the PR output coupler remained plane ($RoC_1 = RoC_2 = \infty$), the L_2 resonant cavity parameter was fixed at 10 mm, and L_1 was optimized in order to achieve the best overlap between the pump and the laser mode volumes for each case. Figure 10 shows the strong influence of the Nd:YAG laser rod diameter on the laser beam profile calculated through LASCAD® analysis. The laser beam profile revealed a top hat shape for the 5.2, 5.3, 5.4, 5.5, and 5.6 mm laser rod diameters. As the rod diameter increased, the laser beam profile showed lower energy in its center, and a doughnut-shaped laser beam profile emerged from the 5.8 mm Nd:YAG laser rod diameter. The L_1 resonant cavity parameter, the $1/e^2$ width beam waist value ω, the laser beam divergence half-angles at the X-Z and Y-Z planes (θ_x and θ_y, respectively), the multimode and TEM_{00} mode laser power, the laser beam quality factors M_x^2 and M_y^2 and the 2D and 3D view of the laser beam profile are also presented in Figure 10 as a function of the laser rod diameter.

Rod diameter (mm)	L_1 resonant cavity parameter (mm)	$1/e^2$ width beam waist ω (μm)	Divergence half-angle Θ_x and Θ_y (°)	Multimode laser power (W)	TEM_{00} mode laser power (W)	Beam quality factors M_x^2 and M_y^2	Beam profile 2D	Beam profile 3D
5.2	248	1150	$\Theta_x = 0.016°$ $\Theta_y = 0.011°$	83.6	36.1	$M_x^2 = 1.88$ $M_y^2 = 1.00$	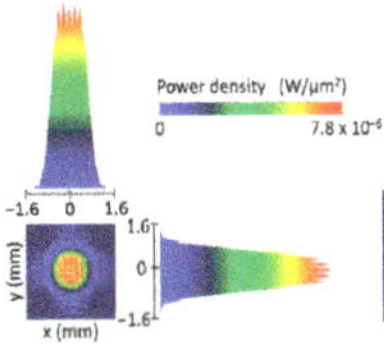	
5.3	252	1188	$\Theta_x = 0.019°$ $\Theta_y = 0.012°$	82.6	37.5	$M_x^2 = 2.30$ $M_y^2 = 1.00$	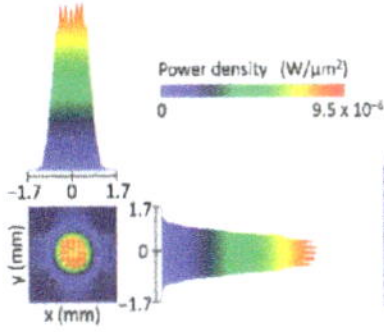	
5.4	257	1245	$\Theta_x = 0.018°$ $\Theta_y = 0.010°$	86.1	37.2	$M_x^2 = 2.83$ $M_y^2 = 1.00$	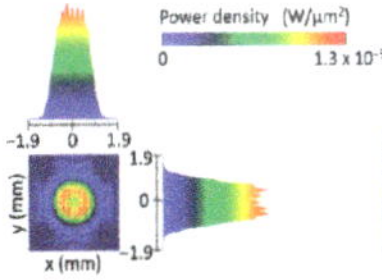	
5.5	260	1233	$\Theta_x = 0.011°$ $\Theta_y = 0.010°$	87.2	37.5	$M_x^2 = 1.25$ $M_y^2 = 1.00$	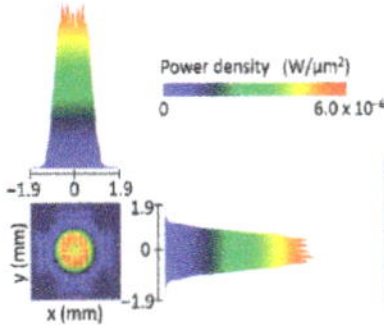	
5.6	264	1215	$\Theta_x = 0.011°$ $\Theta_y = 0.011°$	85.8	36.5	$M_x^2 = 1.00$ $M_y^2 = 1.05$	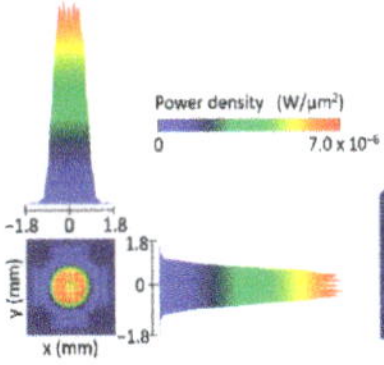	
5.7	269	1245	$\Theta_x = 0.010°$ $\Theta_y = 0.016°$	89.3	34.6	$M_x^2 = 1.00$ $M_y^2 = 2.23$	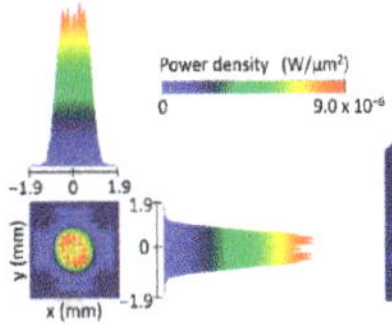	
5.8	274	1375	$\Theta_x = 0.012°$ $\Theta_y = 0.008°$	90.2	39.0	$M_x^2 = 1.90$ $M_y^2 = 1.00$	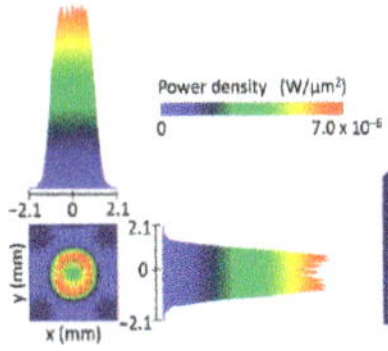	

Figure 10. Summary of the laser beam profiles obtained as a function of the Nd:YAG laser rod diameter through LASCAD® analysis.

Figure 11a presents the 3D view of the top hat laser beam profile obtained from the 5.5 mm diameter laser rod with beam quality factors of $M_x^2 = 1.25$, $M_y^2 = 1.00$, and TEM_{00} mode laser power of 37.5 W. For the doughnut-shaped laser beam profile obtained from the 5.8 mm laser rod diameter, 39.0 W TEM_{00} mode laser power was numerically attained, with beam quality factors of $M_x^2 = 1.90$, $M_y^2 = 1.00$, as shown in Figure 11b.

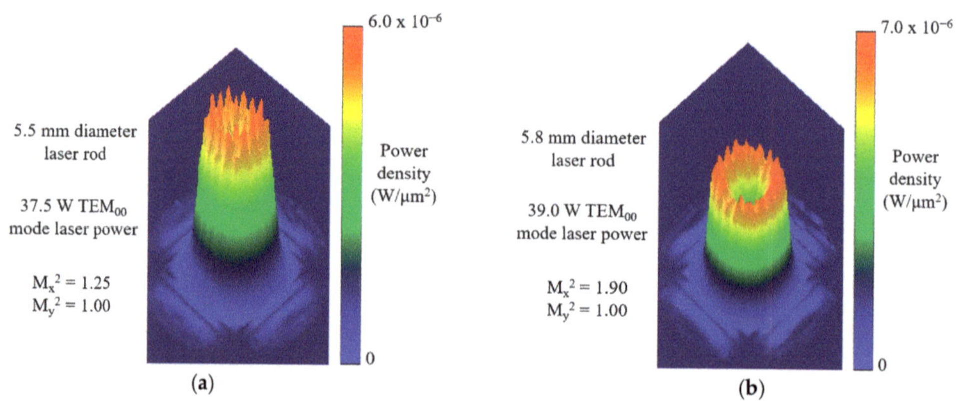

Figure 11. Numerically simulated 3D laser beam profile for (**a**) the 5.5 mm diameter and (**b**) the 5.8 mm diameter laser rods.

4. Discussion

The previously presented results demonstrate, to the best of our knowledge, the first numerical simulation of doughnut-shaped and top hat solar laser beam profiles. The light guides were essential to symmetrically side-pumped the laser rod with six narrow rectangular pump columns, whereas the laser rod diameter was the key parameter to obtain different laser beam profiles. Table 2 establishes the comparison between the present work and the experimental/numerical records for collection and solar-to-TEM_{00}/multimode power conversion efficiencies for the top hat, doughnut-shaped, and multimode solar laser beams. A top hat laser beam profile with 9.4 W/m^2 collection efficiency and 0.99% solar-to-TEM_{00} mode power conversion efficiency was obtained from the 5.5 mm diameter laser rod. As far as we know, this kind of solar laser beam was never reported in experimental essays. A doughnut-shaped laser beam was also numerically attained, from the 5.8 mm diameter laser rod, with 9.8 W/m^2 TEM_{00} mode laser power collection and 1.03% solar-to-TEM_{00} mode power conversion efficiencies, corresponding to an increase of 2.2 and 1.9 times, respectively, compared to the current experimental record for doughnut-shaped solar laser power achieved by Vistas et al. [52]. As experimentally demonstrated by Liang et al. [50], TEM_{01}^* mode laser beam profiles, commonly known as doughnut-shaped, showed only slightly higher laser power than TEM_{00} mode laser beam profiles, regarding similar pumping conditions. Hence, the numerically calculated TEM_{00} mode laser power can be a suitable approximation of the TEM_{01}^* mode laser power. The 90.2 W multimode solar laser power was obtained with the 5.8 mm laser rod diameter, corresponding to 22.6 W/m^2 collection and 2.37% solar-to-multimode power conversion efficiencies, similar to the previous experimental results regarding solar laser side-pumping configurations [50]. Therefore, the novel solar laser configuration numerically demonstrated the ability to produce doughnut-shaped and top hat laser beams, with major importance for the industry, while competing in efficiency with the most advanced solar laser systems so far.

Table 2. Comparison between the present work and the experimental or numerical records for collection and solar-to-TEM_{00}/multimode power conversion efficiencies for the top hat, doughnut-shaped (TEM_{01}^{*} mode), and multimode solar laser beams.

Scheme	Laser Mode	Collection Efficiency (W/m^2)	Conversion Efficiency (%)	Beam Quality Factors
Vistas et al., 2018 [52] (Experimental)	TEM_{01}^{*}	4.5	0.54	Not reported
Liang et al., 2021 [49] (Numerical)	Multimode	23.3	2.82	Not reported
	TEM_{00}	13.7	1.44	$M_x^2 = 1.00$ $M_y^2 = 1.04$
Present work (Numerical)	Multimode	22.6	2.37	Not reported
	TEM_{01}^{*}	9.8	1.03	$M_x^2 = 1.90$ $M_y^2 = 1.00$
	Top hat beam	9.4	0.99	$M_x^2 = 1.25$ $M_y^2 = 1.00$

5. Conclusions

As far as we know, the novel solar laser system reported here successfully accomplished the first numerical analysis on the efficient generation of doughnut-shaped and top hat solar laser beam profiles. The solar energy was firstly collected and concentrated by six Fresnel lenses and then redirected toward six secondary concentrators through six plane folding mirrors. The secondary fused silica concentrators further compressed the solar radiation onto the six twisted fused silica light guides that transformed the incident near-Gaussian energy distribution into a narrow rectangular pump column, allowing a symmetrical side-pumping around the 5.5 mm diameter, 35 mm length Nd:YAG laser rod inside a cylindrical cavity with water cooling. An asymmetric resonator ensured an optimum mode overlap. A top hat laser beam profile ($M_x^2 = 1.25$, $M_y^2 = 1.00$) was computed through both ZEMAX$^{®}$ and LASCAD$^{®}$ analysis, with 9.4 W/m^2 TEM_{00} mode laser power collection and 0.99% solar-to-TEM_{00} mode power conversion efficiencies. By using a 5.8 mm laser rod diameter, a doughnut-shaped solar laser beam profile ($M_x^2 = 1.90$, $M_y^2 = 1.00$) was also produced. The 90.2 W multimode solar laser power was attained with the 5.8 mm laser rod diameter, corresponding to 22.6 W/m^2 collection and 2.37% solar-to-multimode power conversion efficiencies, similar to the previous experimental results regarding solar laser side-pumping configurations [50]. These results demonstrated the laser rod diameter as a key parameter to accomplish top hat and doughnut-shaped laser beams. Even though doughnut-shaped solar laser beams were already experimentally obtained, the methods to produce them were not fully exploited. Hence, the first numerical simulation of doughnut-shaped and top hat solar laser beam profiles reported here significantly contributes to understanding how to generate such beams, particularly the top hat solar laser beam that was not yet demonstrated in practical essays. Future research aims to experimentally validate the results obtained here for both the top hat and doughnut-shaped laser beam profiles. The major barrier to the application of this technology is related to the availability of solar radiation at Earth's surface. Although solar energy can still be collected during cloudy days, the solar laser system's efficiency can significantly drop. However, in space, solar radiation is a more reliable resource, and, as demonstrated here, the difference between these solar laser beams can be useful for future communications systems. Upcoming advancements may expand solar-pumped laser applications into new areas and provide an alternative cost-effective approach to generate these laser beam profiles, essential for high accuracy laser materials processing, particularly at nanoscale.

Author Contributions: Conceptualization, D.L. and M.C.; methodology, M.C., D.L., J.A. and C.R.V.; software, M.C., D.L., C.R.V., H.C. and D.G.; validation, M.C. and D.L.; formal analysis, M.C., D.L., C.R.V. and D.G.; investigation, M.C. and D.L.; resources, D.L. and J.A.; data curation, M.C. and D.L.; writing—original draft preparation, M.C.; writing—review and editing, M.C., D.L., C.R.V., D.G., B.D.T., H.C. and J.A.; visualization, M.C.; supervision, D.L. and C.R.V.; project administration, D.L.; funding acquisition, D.L. All authors have read and agreed to the published version of the manuscript.

Funding: This research was funded by Fundação para a Ciência e a Tecnologia—Ministério da Ciência, Tecnologia e Ensino Superior, grant number UIDB/00068/2020.

Institutional Review Board Statement: Not applicable.

Informed Consent Statement: Not applicable.

Data Availability Statement: Not applicable.

Acknowledgments: The fellowship grants SFRH/BD/145322/2019, PD/BD/142827/2018, PD/BD/128267/2016, SFRH/BPD/125116/2016 and CEECIND/03081/2017 of Miguel Catela, Dario Garcia, Bruno D. Tibúrcio Cláudia R. Vistas and Joana Almeida, respectively, are acknowledged.

Conflicts of Interest: The authors declare no conflict of interest. The funders had no role in the design of the study, in the collection, analyses, or interpretation of data, in the writing of the manuscript, or in the decision to publish the results.

References

1. Liang, D.; Almeida, J. Highly efficient solar-pumped Nd:YAG laser. *Opt. Express* **2011**, *19*, 26399. [CrossRef]
2. Summerer, L.; Purcell, O. Concepts for Wireless Energy Transmission via Laser. *ESA-Adv. Concepts Team* **2008**, 1–10.
3. Hemmati, H.; Biswas, A.; Djordjevic, I.B. Deep-space optical communications: Future perspectives and applications. *Proc. IEEE* **2011**, *99*, 2020–2039. [CrossRef]
4. Lando, M.; Kagan, J.A.; Shimony, Y.; Kalisky, Y.Y.; Noter, Y.; Yogev, A.; Rotman, S.R.; Rosenwaks, S. Solar-pumped solid state laser program. In Proceedings of the 10th Meeting on Optical Engineering in Israel, Tel-Aviv, Israel, 1–6 March 1997; Shaldov, I., Rotman, S.R., Eds.; SPIE: Jerusalem, Israel, 1997; Volume 3110, pp. 196–201.
5. Rather, J.D.G.; Gerry, E.T.; Zeiders, G.W. Investigation of possibilities for solar powered high energy lasers in space. *NASA Tech. Rep. Serv.* **1977**. Available online: https://ntrs.nasa.gov/search.jsp?R=19770019534 (accessed on 13 May 2021).
6. Vasile, M.; Maddock, C.A. Design of a formation of solar pumped lasers for asteroid deflection. *Adv. Sp. Res.* **2012**, *50*, 891–905. [CrossRef]
7. Yabe, T.; Uchida, S.; Ikuta, K.; Yoshida, K.; Baasandash, C.; Mohamed, M.S.; Sakurai, Y.; Ogata, Y.; Tuji, M.; Mori, Y.; et al. Demonstrated fossil-fuel-free energy cycle using magnesium and laser. *Appl. Phys. Lett.* **2006**, *89*, 261107. [CrossRef]
8. Graydon, O. A sunny solution. *Nat. Photonics* **2007**, *1*, 495–496. [CrossRef]
9. Oliveira, M.; Liang, D.; Almeida, J.; Vistas, C.R.; Gonçalves, F.; Martins, R. A path to renewable Mg reduction from MgO by a continuous-wave Cr:Nd:YAG ceramic solar laser. *Sol. Energy Mater. Sol. Cells* **2016**, *155*, 430–435. [CrossRef]
10. Koechner, W. *Solid-State Laser Engineerings*, 6th ed.; Springer: New York, NY, USA, 2006; ISBN 978-0-387-29338-7.
11. Duocastella, M.; Arnold, C.B. Bessel and annular beams for materials processing. *Laser Photonics Rev.* **2012**, *6*, 607–621. [CrossRef]
12. Le, H.; Penchev, P.; Henrottin, A.; Bruneel, D.; Nasrollahi, V.; Ramos-de-Campos, J.A.; Dimov, S. Effects of top-hat laser beam processing and scanning strategies in laser micro-structuring. *Micromachines* **2020**, *11*, 221. [CrossRef]
13. Dickey, F.M.; Holswade, S.C. *Laser Beam Shaping, Theory and Techniques*; Marcel Dekker, Inc.: New York, NY, USA; Basel, Switzerland, 2000; ISBN 9780735405752.
14. Grojean, R.E.; Feldman, D.; Roach, J.F. Production of flat top beam profiles for high energy lasers. *Rev. Sci. Instrum.* **1980**, *51*, 375–376. [CrossRef] [PubMed]
15. Frieden, B.R. Lossless Conversion of a Plane Laser Wave to a Plane Wave of Uniform Irradiance. *Appl. Opt.* **1965**, *4*, 1400. [CrossRef]
16. Hoffnagle, J.A.; Jefferson, C.M. Design and performance of a refractive optical system that converts a Gaussian to a flattop beam. *Appl. Opt.* **2000**, *39*, 5488. [CrossRef]
17. Duerr, F.; Thienpont, H. Refractive laser beam shaping by means of a functional differential equation based design approach. *Opt. Express* **2014**, *22*, 8001. [CrossRef] [PubMed]
18. Leger, J.R.; Chen, D.; Wang, Z. Diffractive optical element for mode shaping of a Nd:YAG laser. *Opt. Lett.* **1994**, *19*, 108. [CrossRef]
19. Dickey, F.M. Gaussian laser beam profile shaping. *Opt. Eng.* **1996**, *35*, 3285. [CrossRef]
20. Račiukaitis, G.; Stankevičius, E.; Gečys, P.; Gedvilas, M.; Bischoff, C.; Jäger, E.; Umhofer, U.; Völklein, F. Laser processing by using diffractive optical laser beam shaping technique. *J. Laser Micro Nanoeng.* **2011**, *6*, 37–43. [CrossRef]
21. Ding, X.Y.; Ren, Y.X.; Lu, R. De Shaping super-Gaussian beam through digital micro-mirror device. *Sci. China Phys. Mech. Astron.* **2015**, *58*, 1–6. [CrossRef]

22. Forbes, A.; Dudley, A.; McLaren, M. Creation and detection of optical modes with spatial light modulators. *Adv. Opt. Photonics* **2016**, *8*, 200. [CrossRef]
23. Litvin, I.A.; Forbes, A. Intra cavity flat top beam generation. *Opt. Express* **2009**, *17*, 15891–15903. [CrossRef]
24. Litvin, I.A.; King, G.; Strauss, H. Beam shaping laser with controllable gain. *Appl. Phys. B Lasers Opt.* **2017**, *123*, 1–5. [CrossRef]
25. Kana, E.T.; Bollanti, S.; Di Lazzaro, P.; Murra, D.; Bouba, O.; Onana, M.B. Laser beam homogenization: Modeling and comparison with experimental results. *Opt. Commun.* **2006**, *264*, 187–192. [CrossRef]
26. Wallace, L. *Efficient Transformation of Gaussian Beams into Uniform, Rectangular Intensity Distributions*; Coherent, Inc.: Santa Clara, CA, USA, 2011.
27. Homburg, O.; Mitra, T. Gaussian-to-top-hat beam shaping: An overview of parameters, methods, and applications. In Proceedings of the Laser Reson. Microresonators Beam Control XIV, San Francisco, CA, USA, 21–26 January 2012; Volume 8236, p. 82360A. [CrossRef]
28. Hracek, B.; Bäuerle, H. New Ways to Generate Flat-Top Profiles. *Opt. Photonik* **2015**, *10*, 16–18. [CrossRef]
29. Santarsiero, M.; Borghi, R. Correspondence between super-Gaussian and flattened Gaussian beams. *J. Opt. Soc. Am. A* **1999**, *16*, 188–190. [CrossRef]
30. Shealy, D.L.; Hoffnagle, J.A. Laser beam shaping profiles and propagation. *Appl. Opt.* **2006**, *45*, 5118–5131. [CrossRef]
31. Kaszei, K.; Boone, C. Flat-Top Laser Beams: Their Uses and Benefits. Available online: https://www.laserfocusworld.com/optics/article/14197298/flattop-laser-beams-their-uses-and-benefits (accessed on 17 May 2021).
32. Fuse, K. Beam Shaping for Advanced Laser Materials Processing. *Laser Tech. J.* **2015**, *12*, 19–22. [CrossRef]
33. Hung, Y.-J.; Chang, H.-J.; Chang, P.-C.; Lin, J.-J.; Kao, T.-C. Employing refractive beam shaping in a Lloyd's interference lithography system for uniform periodic nanostructure formation. *J. Vac. Sci. Technol. B Nanotechnol. Microelectron. Mater. Process. Meas. Phenom.* **2017**, *35*, 030601. [CrossRef]
34. Ramos, S.; Stork, W.; Heussner, N. Multiple-pulse damage thresholds of retinal explants using top hat profiles. In Proceedings of the Optical Interactions with Tissue and Cells XXXI, San Francisco, CA, USA, 1–2 February 2020; Volume 11238, p. 112380D. [CrossRef]
35. Zhang, Y.; Wu, C.; Song, Y.; Si, K.; Zheng, Y.; Hu, L.; Chen, J.; Tang, L.; Gong, W. Machine learning based adaptive optics for doughnut-shaped beam. *Opt. Express* **2019**, *27*, 16871. [CrossRef]
36. Tumkur, T.U.; Voisin, T.; Shi, R.; Depond, P.J.; Roehling, T.T.; Wu, S.; Crumb, M.F.; Roehling, J.D.; Guss, G.; Khairallah, S.A.; et al. Nondiffractive beam shaping for enhanced optothermal control in metal additive manufacturing. *Sci. Adv.* **2021**, *7*, 9358–9373. [CrossRef]
37. Kiss, Z.J.; Lewis, H.R.; Duncan, R.C. Sun pumped continuous optical maser. *Appl. Phys. Lett.* **1963**, *2*, 93–94. [CrossRef]
38. Maiman, T.H. Stimulated optical radiation in Ruby. *Nature* **1960**, *187*, 493–494. [CrossRef]
39. Arashi, H.; Oka, Y.; Sasahara, N.; Kaimai, A.; Ishigame, M. A Solar-Pumped cw 18 W Nd:YAG Laser. *Jpn. J. Appl. Phys.* **1984**, *23*, 1051–1053. [CrossRef]
40. Weksler, M.; Shwartz, J. Solar-pumped solid-state lasers. *IEEE J. Quantum Electron.* **1988**, *24*, 1222–1228. [CrossRef]
41. Lando, M.; Kagan, J.; Linyekin, B.; Dobrusin, V. A solar-pumped Nd:YAG laser in the high collection efficiency regime. *Opt. Commun.* **2003**, *222*, 371–381. [CrossRef]
42. Liang, D.; Almeida, J.; Vistas, C.R.; Guillot, E. Solar-pumped Nd:YAG laser with 31.5 W/m² multimode and 7.9 W/m² TEM$_{00}$-mode collection efficiencies. *Sol. Energy Mater. Sol. Cells* **2017**, *159*, 435–439. [CrossRef]
43. Liang, D.; Vistas, C.R.; Tibúrcio, B.D.; Almeida, J. Solar-pumped Cr:Nd:YAG ceramic laser with 6.7% slope efficiency. *Sol. Energy Mater. Sol. Cells* **2018**, *185*, 75–79. [CrossRef]
44. Yabe, T.; Ohkubo, T.; Uchida, S.; Yoshida, K.; Nakatsuka, M.; Funatsu, T.; Mabuti, A.; Oyama, A.; Nakagawa, K.; Oishi, T.; et al. High-efficiency and economical solar-energy-pumped laser with Fresnel lens and chromium codoped laser medium. *Appl. Phys. Lett.* **2007**, *90*, 261120. [CrossRef]
45. Dinh, T.H.; Ohkubo, T.; Yabe, T.; Kuboyama, H. 120 watt continuous wave solar-pumped laser with a liquid light-guide lens and an Nd:YAG rod. *Opt. Lett.* **2012**, *37*, 2670–2672. [CrossRef]
46. Liang, D.; Almeida, J. Solar-Pumped TEM$_{00}$ Mode Nd:YAG laser. *Opt. Express* **2013**, *21*, 25107. [CrossRef] [PubMed]
47. Xu, P.; Yang, S.; Zhao, C.; Guan, Z.; Wang, H.; Zhang, Y.; Zhang, H.; He, T. High-efficiency solar-pumped laser with a grooved Nd:YAG rod. *Appl. Opt.* **2014**, *53*, 3941–3944. [CrossRef]
48. Guan, Z.; Zhao, C.; Li, J.; He, D.; Zhang, H. 32.1 W/m² continuous wave solar-pumped laser with a bonding Nd:YAG/YAG rod and a Fresnel lens. *Opt. Laser Technol.* **2018**, *107*, 158–161. [CrossRef]
49. Liang, D.; Almeida, J.; Tibúrcio, B.; Catela, M.; Garcia, D.; Costa, H.; Vistas, C. Seven-rod pumping approach for the most efficient production of TEM$_{00}$-mode solar laser power by a Fresnel lens. *J. Sol. Energy Eng.* **2021**, *143*, 1–28. [CrossRef]
50. Liang, D.; Vistas, C.R.; Almeida, J.; Tibúrcio, B.D.; Garcia, D. Side-pumped continuous-wave Nd:YAG solar laser with 5.4% slope efficiency. *Sol. Energy Mater. Sol. Cells* **2019**, *192*, 147–153. [CrossRef]
51. Almeida, J.; Liang, D.; Vistas, C.R. A doughnut-shaped Nd:YAG solar laser beam. *Opt. Laser Technol.* **2018**, *106*, 1–6. [CrossRef]
52. Vistas, C.R.; Liang, D.; Almeida, J.; Tibúrcio, B.D.; Garcia, D. A doughnut-shaped Nd:YAG solar laser beam with 4.5 W/m² collection efficiency. *Sol. Energy* **2019**, *182*, 42–47. [CrossRef]
53. Rioux, M.; Tremblay, R.; Bélanger, P.A. Linear, annular, and radial focusing with axicons and applications to laser machining. *Appl. Opt.* **1978**, *17*, 1532. [CrossRef]

54. Cherezova, T.; Chesnokov, S.; Kaptsov, L.; Kudryashov, A. Doughnut-like laser beam output formation by intracavity flexible controlled mirror. *Opt. Express* **1998**, *3*, 180. [CrossRef]
55. Litvin, I.A.; Ngcobo, S.; Naidoo, D.; Ait-Ameur, K.; Forbes, A. Doughnut laser beam as an incoherent superposition of two petal beams. *Opt. Lett.* **2014**, *39*, 704. [CrossRef]
56. Bernardes, P.H.; Liang, D. Solid-state laser pumping by light guides. *Appl. Opt.* **2006**, *45*, 3811–3816. [CrossRef]
57. Lando, M.; Shimony, Y.; Benmair, R.M.J.; Abramovich, D.; Krupkin, V.; Yogev, A. Visible solar-pumped lasers. *Opt. Mater.* **1999**, *13*, 111–115. [CrossRef]
58. Liang, D.; Almeida, J.; Garcia, D. Comparative Study of Cr:Nd:YAG and Nd:YAG Solar Laser Performances. In Proceedings of the 8th Iberoamerican Optics Meeting and 11th Latin American Meeting on Optics, Lasers, and Applications, Porto, Portugal, 22–26 July 2013; Volume 8785, p. 87859Y.
59. Vistas, C.R.; Liang, D.; Almeida, J.; Tibúrcio, B.D.; Garcia, D.; Catela, M.; Costa, H.; Guillot, E. Ce:Nd:YAG side-pumped solar laser. *J. Photonics Energy* **2021**, *11*, 1–9. [CrossRef]
60. Lupei, V.; Lupei, A.; Gheorghe, C.; Ikesue, A. Emission sensitization processes involving Nd3+ in YAG. *J. Lumin.* **2016**, *170*, 594–601. [CrossRef]
61. Zhao, B.; Zhao, C.; He, J.; Yang, S. The study of active medium for solar-pumped solid-state lasers. *Acta Opt. Sin.* **2007**, *27*, 1797–1801.
62. ASTM International. *ASTM Standard G173-03, Standard Tables for Reference Solar Spectral Irradiances: Direct Normal and Hemispherical on 37 Tilted Surface*; ASTM: West Conshohocken, PS, USA, 2012.
63. Liang, D.; Almeida, J.; Vistas, C.R.; Guillot, E. Solar-pumped TEM$_{00}$ mode Nd:YAG laser by a heliostat—Parabolic mirror system. *Sol. Energy Mater. Sol. Cells* **2015**, *134*, 305–308. [CrossRef]

Article

Zigzag Multirod Laser Beam Merging Approach for Brighter TEM$_{00}$-Mode Solar Laser Emission from a Megawatt Solar Furnace

Hugo Costa, Joana Almeida, Dawei Liang *, Miguel Catela, Dário Garcia, Bruno D. Tibúrcio and Cláudia R. Vistas

Centro de Física e Investigação Tecnológica (CEFITEC), Departamento de Física, Faculdade de Ciências e Tecnologia, Universidade NOVA de Lisboa, 2829-516 Caparica, Portugal; hf.costa@campus.fct.unl.pt (H.C.); jla@fct.unl.pt (J.A.); m.catela@campus.fct.unl.pt (M.C.); kongming.dario@gmail.com (D.G.); brunotiburcio78@gmail.com (B.D.T.); carvistas@gmail.com (C.R.V.)
* Correspondence: dl@fct.unl.pt

Abstract: An alternative multirod solar laser end-side-pumping concept, based on the megawatt solar furnace in France, is proposed to significantly improve the TEM$_{00}$-mode solar laser output power level and its beam brightness through a novel zigzag beam merging technique. A solar flux homogenizer was used to deliver nearly the same pump power to multiple core-doped Nd:YAG laser rods within a water-cooled pump cavity through a fused silica window. Compared to the previous multibeam solar laser station concepts for the same solar furnace, the present approach can allow the production of high-power TEM$_{00}$-mode solar laser beams with high beam brightness. An average of 1.06 W TEM$_{00}$-mode laser power was numerically extracted from each of 1657 rods, resulting in a total of 1.8 kW. More importantly, by mounting 399 rods at a 30° angle of inclination and employing the beam merging technique, a maximum of 5.2 kW total TEM$_{00}$-mode laser power was numerically extracted from 37 laser beams, averaging 141 W from each merged beam. The highest solar laser beam brightness figure of merit achieved was 148 W, corresponding to an improvement of 23 times in relation to the previous experimental record.

Keywords: beam merging; multirod; Nd:YAG; solar furnace; solar flux homogenizer; solar laser; TEM$_{00}$-mode

Citation: Costa, H.; Almeida, J.; Liang, D.; Catela, M.; Garcia, D.; Tibúrcio, B.D.; Vistas, C.R. Zigzag Multirod Laser Beam Merging Approach for Brighter TEM$_{00}$-Mode Solar Laser Emission from a Megawatt Solar Furnace. *Energies* **2021**, *14*, 5437. https://doi.org/10.3390/en14175437

Academic Editor: Bashir A. Arima

Received: 20 July 2021
Accepted: 30 August 2021
Published: 1 September 2021

Publisher's Note: MDPI stays neutral with regard to jurisdictional claims in published maps and institutional affiliations.

1. Introduction

The production of coherent and narrowband laser radiation from broadband solar radiation has gained an ever-increasing importance over the years, providing cost-effective solutions to laser radiation in a more sustainable way. Powered by the largest and most exploitable renewable energy resource, solar-pumped laser systems may offer design simplicity and dismissal of artificial power generation and conditioning equipment. They are especially suited for space-based applications wherein extended run times are required and compactness, reliability, and efficiency are critical, including wireless power transmission [1], deep-space optical communications [2], laser propulsion [3], and asteroid deflection [4]. In addition, its use can be of value for terrestrial applications, in fields such as thermochemistry and materials processing [5–7].

Although the first report of a solar-pumped maser occurred in the 1960s [8], it was only in the 21st century that the most significant progress in solar laser research was made, when aiming to boost the collection efficiency [9–16]—defined as solar laser output power per unit of primary concentrator area [9]. The current record is 32.5 W/m^2, reached through end-side-pumping a single Cr:Nd:YAG ceramic rod with 4.5 mm diameter and 35 mm length, in 2018 [15]. Brightness figure of merit—given by the quotient between the output laser power and the product of the beam quality factors M_x^2 and M_y^2 [9]—is another key parameter for evaluation of the laser performance. In 2017, using a single 4-mm diameter, 35-mm length Nd:YAG rod, the experimental record of 6.46 W brightness figure of merit was attained, as well as a 7.9 W/m^2 TEM$_{00}$-mode collection efficiency [16]. However, the exposure of a single rod to a large amount of pump power can induce excessive heat within it and, consequently, substantial thermal stress effects [17]. These effects are conducive to laser beam distortion, depolarization loss, and fracture of the laser rod, which occurs when the stress originated from the temperature gradients in the laser material surpasses its tensile length [18]. Moreover, overcoming this hindrance is of paramount importance when scaling the laser power without hampering its beam quality [19]. The reduction of the rod diameter makes it more thermally resistant and can improve considerably the beam quality, even though the laser power production could then be severely limited.

Nd:YAG is the most commonly used laser material for highly intense solar pumping in light of the appropriate spectroscopic properties of the dopant [20], and the excellent thermomechanical properties of the host material, giving it great resilience and durability [21]. Its availability and low cost compared to alternative laser materials are also factors that are taken into consideration. The Nd:YAG laser is a four-level system, as depicted by the simplified energy level diagram in Figure 1. The laser beams are typically emitted at a wavelength of 1064 nm, originated from the transition from the $^4F_{3/2}$ to the $^4I_{11/2}$ levels [18]. In a four-level laser medium, the lower laser level is well above the ground state, being quickly depopulated by multiple phonon transitions. This means that reabsorption of the laser radiation can be avoided, and a lower threshold pump power can be achieved, making it easier to obtain stimulated emission with relatively low pump power.

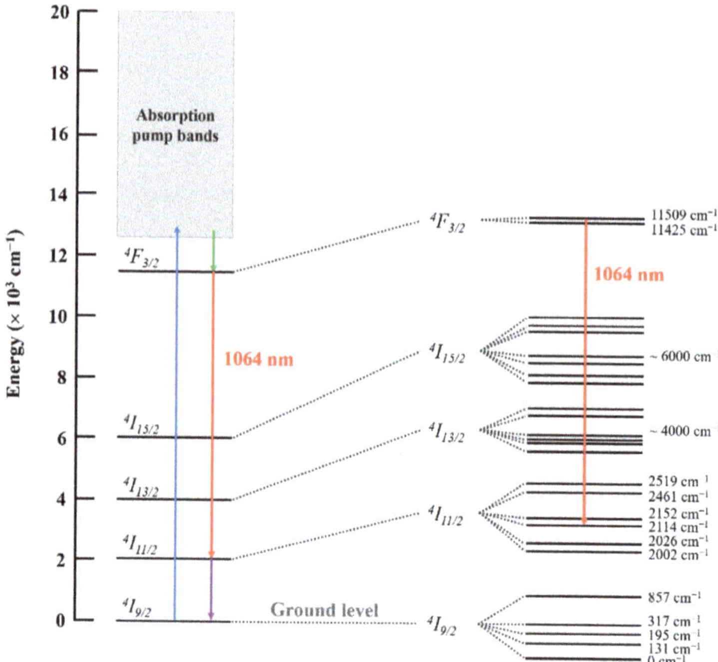

Figure 1. Simplified energy level diagram of Nd:YAG (based on [18]).

Core-doped ceramic media, which are laser active only in their core, possess a cladding around it with the same properties as the host material. The cladding can be either undoped or doped with a different material that efficiently absorbs light at the signal wavelength. This technology arose with the purpose of enhancing the brightness of diode-pumped solid-state lasers, compared to that with conventional single-crystal laser media [22,23], and has shown to be promising for solar laser research [24–27]. Since the cross-section of the laser rod is widened by the cladding, the laser active region has room for wider Gaussian intensity distributions without truncating its wings. The upper laser level becomes even more populated than the lower level with the increase of the average intensity in the doped part of the laser rod, and a more efficient built-up inversion can be obtained. Therefore, the higher the population inversion, the more efficient the laser power production is. The impact of diffraction losses on the laser beam quality may also be diminished [22,23].

As previously mentioned, lasers are indispensable materials-processing tools, as they enable the productivity enhancement that manufacturers look for to extend their business to new segments. In recent times, this search has resulted in the simultaneous emission of several laser beams on a single workpiece, where each one could be optimized to perform only a part of the overall process [28]. It has proven its effectiveness in applications such as welding, brazing, surface texturing [28], microstructuring [29], laser ablation, drilling, and cutting [30,31], achieving more positive outcomes compared to that with the conventional single laser beam. In similar fashion, researchers have been proposing concepts based on the distribution of concentrated radiation among multiple rods [32–39] to abate the thermal stress effects. Through numerical work, multirod laser stations have shown great potential to be used in megawatt solar furnaces (MWSF) [32,36,38]. For the MWSF of Procédés, Matériaux et Energie Solaire—Centre National de la Recherche Scientifique (PROMES-CNRS), total multimode solar laser power of 22.84 kW was numerically attained from a side-pumping laser system with a solar flux homogenizer and 12 Nd:YAG rods of 10 mm diameter and 516 mm length, resulting in 12.48 W/m^2 collection and 2.28% solar-to-laser power conversion efficiencies [36]. This concept, however, displayed strong thermal lensing in each rod, impairing the laser beam quality. In consideration of this, a 32-rod side-pumping scheme was proposed for that same facility [38]. With rods of 6 mm diameter and eight different lengths, it showed a considerable alleviation of the thermal lensing effects and improvement of the laser beam quality factors, possibly enabling the generation of 9.44 kW total laser power from 32 laser beams with quasi-Gaussian profiles. Nevertheless, in both these schemes, the thermal lensing was still very pronounced in the large diameter rods, so operating with rods with smaller diameter was not possible. The maximum extraction of TEM_{00}-mode solar laser is only conceivable by adopting laser rods of small diameter and resonator configurations of large-mode volume, which leads to a substantial improvement of the laser beam quality [16,40–44]. Gaussian TEM_{00}-mode laser beams are desired for many applications due to having the lowest divergence and, consequently, the highest intensity and brightness, allowing greater energy concentration for long distances [18]. This type of laser beam is typically used in materials processing as a result of the preservation of its shape when passing through an optical system consisting of multiple elements employed to guide the laser radiation to the workpiece [45]. These features, coupled with the possibility to be focused to a diffraction-limited spot, make TEM_{00}-mode laser beams the best option for microfabrication applications [45], for example. Moreover, by making use of a laser beam merging technique, TEM_{00}-mode laser beams of higher power, enhanced beam quality and, thus, higher brightness can be obtained [46].

In that regard, an alternative multirod end-side-pumping concept is proposed to significantly improve the TEM_{00}-mode solar laser output power level and its beam brightness through a novel zigzag multirod laser beam merging technique. It comprised of the MWSF of PROMES-CNRS for collection and concentration of solar light, a solar flux homogenizer with hexagonal shaped faces, and multiple arrays of core-doped Nd-YAG rods placed after a fused silica window at the homogenizer's output face. Surrounding these rods is a hexagonal shaped pump cavity, used to help in providing approximately the same amount of solar power to pump each rod. All of the design parameters of the present approach were numerically optimized on both Zemax® nonsequential ray-tracing and laser cavity analysis and design (LASCAD™) software.

2. Description of the Multibeam Solar Laser Station

2.1. Megawatt Solar Furnace of PROMES-CNRS

The MWSF of the PROMES-CNRS laboratory (Figure 2), in Odeillo, France, is comprised of a 63-heliostat field with a total surface area of 2835 m^2 and a 1830 m^2 facetted truncated parabolic mirror with 18 m focal length, mounted on the north face of an eight-story building [47]. The 63 heliostats are distributed among eight flat terraces at different heights, placed in staggered rows (Figure 2c). Each heliostat tracks the Sun, collects the solar rays and redirects them horizontally toward the parabolic mirror. It then focuses 1 MW of solar power into an 80 cm diameter Gaussian spot, reaching a peak flux value beyond 10 W/mm^2 [48]. The light source in Zemax® was programmed accordingly for the design and optimization of the multirod end-side-pumping concept, by manually adjusting the G_X and G_Y parameters, which represent the Gaussian distribution parameters in the x- and y-axes. The desired light distribution at the focus was attained at $G_X = G_Y = 29{,}000$.

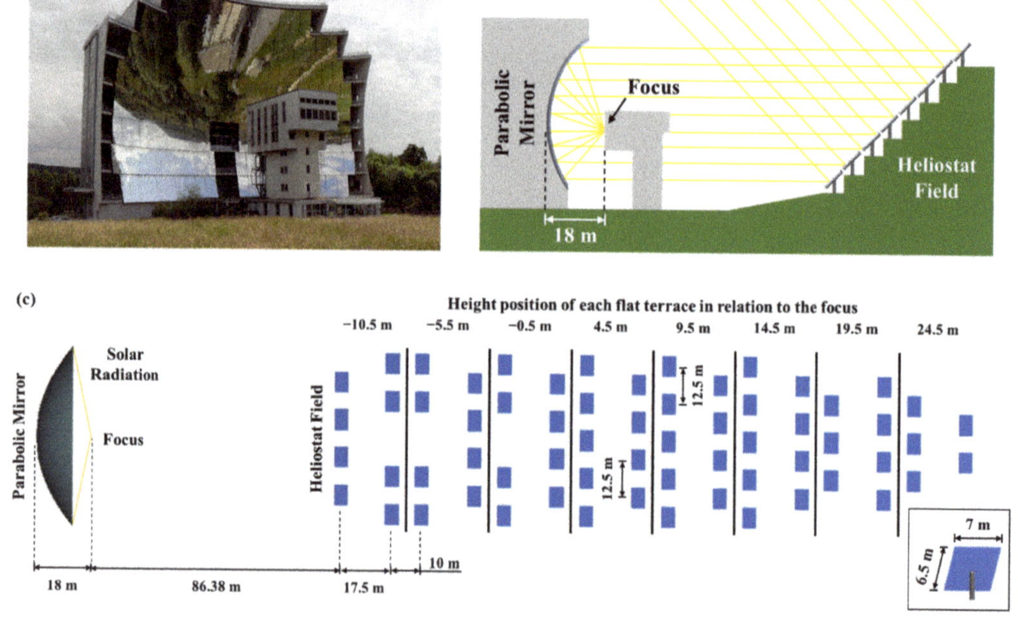

Figure 2. (a) Photograph and (b) simplified schematic of the MWSF facility in Odeillo, France. (c) Top-view of the solar energy collection and concentration system (based on [47]). The dimensions of a heliostat are presented in the inset of (c).

2.2. Solar Flux Homogenizer and Arrays of Core-Doped Nd:YAG Rods

The concentrated solar radiation from the parabolic mirror was received by a hollow homogenizer, with hexagonal input and output faces, placed at the focal zone (Figure 3a). Its role was to reshape the Gaussian solar pump light distribution at its input face into a nearly uniform one at its output end. Both end faces of the homogenizer were designed to present an apothem of 203 mm, while a 390 mm length was necessary to attain the desired type of solar pump light distribution at the exit. An inner wall reflectivity of 98% and an active back-face water cooling of the homogenizer were considered.

As depicted in Figure 3a, at the homogenizer's input face, the concentrated solar radiation presented a Gaussian distribution with a 10.21 W/mm^2 peak flux. The solar rays underwent multiple reflections along its length, and a nearly uniform profile with an average flux of 4.38 W/mm^2 was achieved at the output end, after which a thin hexagonal fused silica window was positioned. At the input/output face of the homogenizer, a solar rays' rim angle of nearly 80° was also found.

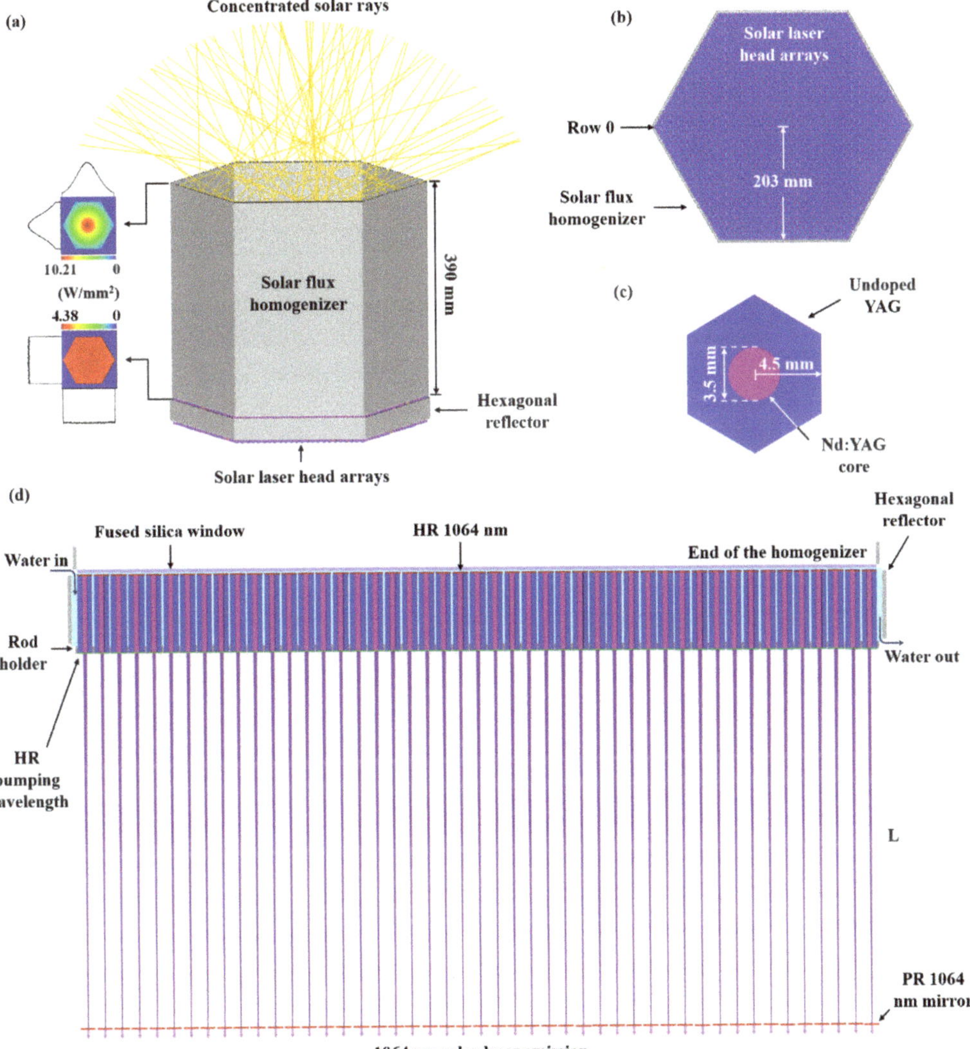

Figure 3. (**a**) 3D and (**b**) top view of the solar flux homogenizer, 1657 laser rods and the hexagonal reflector. The concentrated light distribution at the input and output faces of the homogenizer is also presented in (**a**). (**c**) Top view of a core-doped Nd:YAG rod. (**d**) Cross-sectional view of the lower part of the system, showing the central row (row 0, represented in (**b**)) with 47 solar laser rods, and the extraction of a laser beam from each one. L represents the separation length between the highly reflective (HR) at pumping wavelength coating on the lower end face of the rod and the partially reflective (PR) at 1064 nm output mirror.

Below the fused silica window, 1657 core-doped Nd:YAG laser rods were mounted vertically (Figure 3b,d), totaling 47 rows with different number of rods. Each rod had a 45 mm length and was composed of a cylindrical Nd:YAG core with a 3.5 mm diameter, and a hexagonal undoped YAG cladding with a 4.5 mm apothem (Figure 3c). With both homogenizer and undoped YAG claddings being hexagonal shaped, the rods could be uniformly tiled on a flat plane with minimal gap losses. Surrounding the rods was a hexagonal reflector with a 206 mm apothem, 39 mm length, and 98% reflectivity for its inner walls. A single rod holder was used for mechanical fixation of the lower end of all core-doped Nd:YAG rods. The rods, the lower face of the fused silica window and the inner walls of the hexagonal reflector were water cooled, entering from the space between the fused silica window and the reflector, and exiting from the one between the rod holder and that same reflector. The cooling water helped in handling the temperature regulation because of its high thermal conductivity, as well as specific heat and low viscosity.

For the extraction of a laser beam, a highly reflective (HR) at the laser emission wavelength (1064 nm) coating was added to the top surface of each core-doped Nd:YAG rod. This coating, the laser rod and a partially reflective (PR) at 1064 nm output mirror constituted the laser resonator. The lower surface of each laser rod had a HR at pumping wavelength coating to help maximize the pump power.

3. Numerical Modeling of the Multibeam Solar Laser Station

3.1. Modeling of the Optical Design Parameters through Zemax® Software

For the optimization of the design parameters in Zemax®, a 1000 W/m² solar irradiance, typical during clear sunny days in Odeillo, France, was considered in the analysis. 22 peak absorption wavelengths for the 1.0 at. % Nd:YAG medium were programmed and the spectral irradiance values at each one were used as reference data for the light source, after consulting the standard solar spectrum for AM1.5 [49]. These spectral peaks were positioned at 527 nm, 531 nm, 568 nm, 578 nm, 586 nm, 592 nm, 732 nm, 736 nm, 743 nm, 746 nm, 753 nm, 758 nm, 790 nm, 793 nm, 803 nm, 805 nm, 808 nm, 811 nm, 815 nm, 820 nm, 865 nm, and 880 nm. Each peak wavelength and the corresponding absorption coefficient were also added to the glass catalog data of Zemax® for the Nd:YAG material. The parameters of the light source were first adjusted until a Gaussian profile, identical to that from the MWSF [48], was obtained at the focal spot (inset of Figure 3a). After that, the 16% overlap between the absorption spectrum of the Nd:YAG medium and the solar spectrum [50] was considered when estimating its effective pump power for the numerical analysis of the absorbed pump flux power and distribution within the core-doped Nd:YAG rods. Furthermore, the absorption spectra and wavelength-dependent refractive indices of fused silica and water were added to the glass catalog data of Zemax® to account for absorption losses in those media.

For the analysis of the numerical data within each rod, a detector volume, divided into several voxels, was necessary. The absorbed pump power was obtained through the sum of the power from each individual voxel. The number of analysis rays and voxels were tuned to acquire more accurate results and better image resolution of the detector. Figure 4 shows the absorbed pump flux distribution profiles in the longitudinal and four transversal cross-sections of one of the core-doped Nd:YAG laser media. The red color represents maximum pump flux of 0.25 W/mm³, while blue is attributed to the laser rod regions where there is little or no absorption.

Due to the presence of the solar flux homogenizer, each core-doped rod exhibited absorbed pump flux distribution profiles similar to the ones in Figure 4, with the highest values being detected near the upper end face of the rod. Two regions of higher absorbed pump flux are present here, with a significant decline occurring between them as a result of the refraction of solar rays at the water/YAG and YAG/Nd:YAG interfaces. In addition, an increase of the absorbed pump flux with the distance to the rod's optical axis can be observed in the transversal cross-section profiles.

The absorbed pump flux data obtained from Zemax® were then processed in the LASCAD™ software in order to optimize the solar laser output power associated with the resonator beam parameters. The design parameters of the scheme were further optimized in Zemax® based on these results.

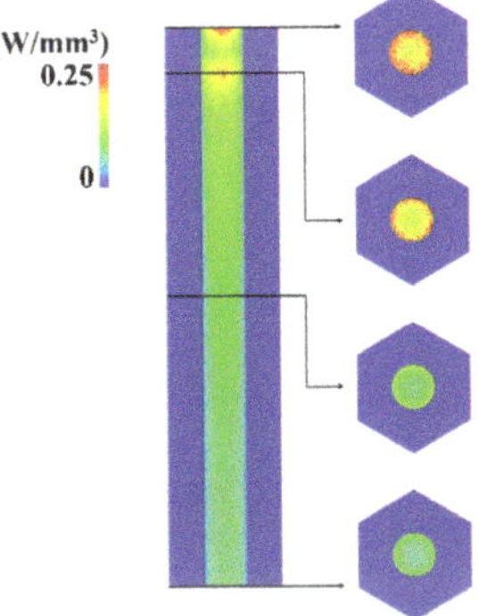

Figure 4. Absorbed pump flux distribution in the longitudinal and four transversal cross-sections of one core-doped Nd:YAG rod.

3.2. Optimization of the TEM$_{00}$-Mode Laser Extraction from Each Core-Doped Nd:YAG Medium through LASCAD™ Software

For the LASCAD™ analysis, a 2.8×10^{-19} cm^2 stimulated emission cross-section, a 230 µs fluorescence lifetime [18], and a 0.003 cm^{-1} typical absorption and scattering loss for the 1.0 at. % Nd:YAG medium were implemented. Moreover, the mean absorbed and intensity-weighted solar pump wavelength of 660 nm was considered [21].

The thermal induced effects in the 3.5 mm diameter, 45 mm length Nd:YAG cores were numerically analyzed in LASCAD™. Figure 5 shows an example of the thermal performance of one of the Nd:YAG cores. Maximum heat load of 0.14 W/mm^3 was attained, as well as a temperature of 306.2 K. A 8.87 N/mm^2 maximum stress intensity was also observed, which is significantly lower than the Nd:YAG material's stress fracture limit of about 200 N/mm^2 [51]. Therefore, the rods would perform relatively well under highly intense solar pumping.

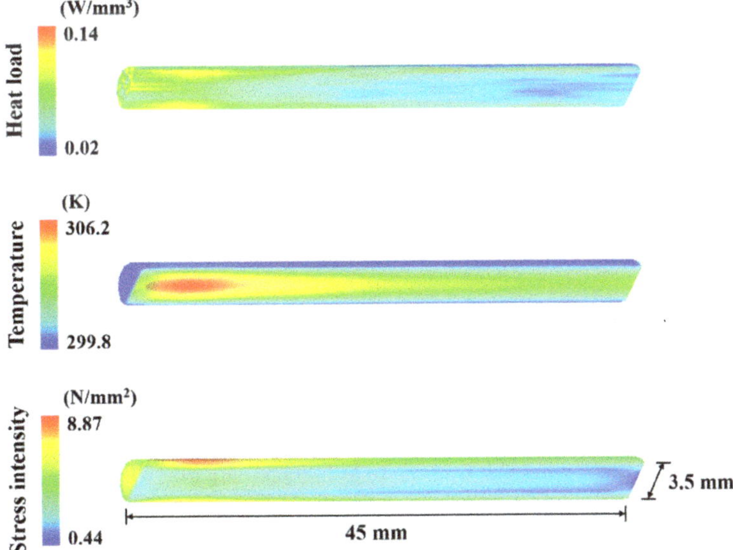

Figure 5. LASCAD™ analysis of heat load, temperature, and stress intensity distribution in one of the Nd:YAG cores.

Each laser resonator was composed of two opposing mirrors, representing the HR 1064 nm coating and the PR 1064 nm output mirror, whose optical axes were aligned with that of the laser rod. A PR mirror reflectivity of 97% provided the maximum TEM$_{00}$-mode laser power.

When the pump power exerts strong influence on the laser rod's thermal lens, the resonator length plays a crucial role in maximizing the extraction of TEM$_{00}$-mode laser beams with high brightness. In a resonator, the smallest size and divergence of a laser beam is obtained by its fundamental transverse mode [18]. For short resonators, the TEM$_{00}$-mode poorly matches the active region and laser oscillates in several modes, which leads to high M^2 factors. With the increase of the resonator length, the size of the TEM$_{00}$-mode beam within the rod also increases, as well as diffraction losses at the rod edges. Consequently, higher order modes can be eliminated. A long resonator was then chosen, to the detriment of laser power, so that only the fundamental mode could oscillate, improving the beam quality and, thus, its brightness [18]. To facilitate the oscillation of TEM$_{00}$-mode, a laser rod of small diameter had to be used since it behaves as an aperture. The separation length and the radius of curvature (RoC) of the PR mirror were also optimized. Figure 6 depicts the laser resonator design for one of the Nd:YAG cores to extract a TEM$_{00}$-mode laser beam.

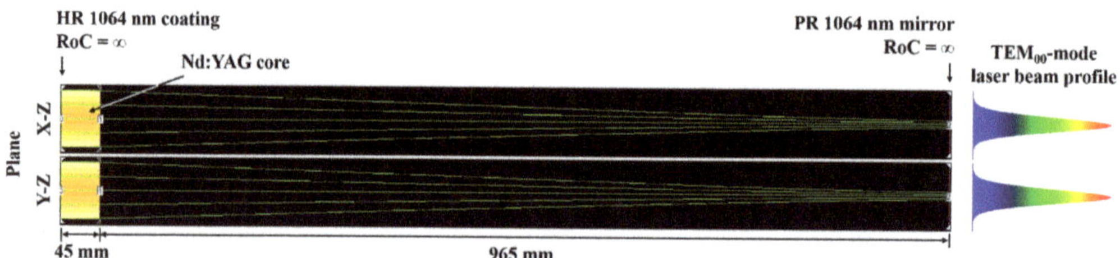

Figure 6. Laser resonator design in LASCAD™ for the extraction of a TEM$_{00}$-mode laser beam from one 3.5 mm diameter, 45 mm length Nd: YAG core. The beam profile is also presented.

In this end-side-pumping scheme, the thermal lensing effects were much less noticeable than in the previous 32-rod side-pumping concept with 6-mm diameter rods for the same solar furnace [38]. This made the use of thinner rods possible, which were pivotal in maximizing the TEM$_{00}$-mode laser power extraction with much longer resonant cavities. From each 3.5 mm diameter, 45 mm length Nd:YAG core, an average of 1.06 W TEM$_{00}$-mode solar laser power was numerically extracted, resulting in a total of about 1.8 kW. Each TEM$_{00}$-mode laser beam displayed good quality factors ($M_x^2 = 1.29$, $M_y^2 = 1.02$), albeit attaining a brightness of only 0.81 W. Total laser power of 15.2 kW in multimode regime was also numerically attained, with an average of 9.18 W from each individual rod, by reducing the resonant cavity length to 50 mm.

4. Extraction of Multiple TEM$_{00}$-Mode Laser Beams with Higher Brightness

By inclining each rod and positioning them at a certain distance from the fused silica window, a zigzag beam merging technique can be implemented to enable the extraction of brighter TEM$_{00}$-mode laser beams, despite having to reduce the number of rods. The angle of inclination of the rods should not be too low to avoid high laser beam transmission losses in the cooling water, since they would have to be positioned at a large distance from the fused silica window.

At approximately 7.62 mm below the fused silica window, 399 core-doped Nd:YAG laser rods were mounted at a 30° angle of inclination (Figure 7), which was the angle that also provided the highest absorbed pump power for each rod. This totals 35 rows with different number of rods, occupying almost the same space as the previous case of the laser beam extraction from each single rod. A wider hexagonal reflector surrounding the rods was adopted, with an apothem of 230 mm and length of 38.34 mm.

To extract a single laser beam from an array of core-doped rods, the resonant cavity was composed of a HR 1064 nm mirror, a certain number of rods (N), the fused silica window with a HR 1064 nm coating on its lower face, a set of horizontal HR 1064 nm folding mirrors (the amount of which differ depending on N), and a PR 1064 nm output mirror. The optical axes of the HR and PR 1064 nm mirrors were aligned with those of the outmost laser rods, while each horizontal HR 1064 nm folding mirror was placed between the lower end faces of two adjacent rods. Furthermore, in this case, the upper end face of each rod presented an antireflective (AR) at 1064 nm coating to allow the transmission of the laser beam toward the HR 1064 nm coating on the lower face of the fused silica window. The latter and the horizontal HR 1064 nm mirrors were essential in folding the laser light from one rod to the next one.

Considering the central row of the solar laser head as row 0, arrays with different N from this row were tested in order to determine which N led to the highest TEM_{00}-mode laser power production. For 3.5 mm diameter, 45 mm length Nd:YAG cores, N = 12 was the optimal number of rods, possibly generating 157.14 W of TEM_{00}-mode laser power. For this reason, N = 12 was used in every row that had at least this number of rods. For rows -17 to -9 and 9 to 17, the maximum N possible was used, varying from N = 8 to 11 rods. From the 399 rods, only 23 remained, positioned in rows -6 to 6, so they were split into two arrays of N = 10 and 13, employing a similar procedure for the extraction of the merged laser to those illustrated in Figure 7.

For the numerical analysis of the merged laser beam performance, a new method was adopted to acquire the text file with the absorbed pump flux/volume data to be imported into LASCAD™, because this software only allows the analysis of a single file. The following steps (Figure 8) were taken:

1. In Zemax®, a long detector volume was added for each of the N Nd:YAG cores of the array. Its length (L_{DET}) was N $\times$ L_{ROD}, with L_{ROD} being the rod length. The number of pixels in the z-axis was N $\times$ P_Z, with P_Z being the number of pixels from the previous case of laser beam extraction from each rod. Each Nd:YAG core occupied a different portion of the corresponding detector.
2. After running the Zemax® simulation, N text files were exported, with the absorbed pump flux/volume data from each detector volume.
3. Tables with information about each z-plane and voxel are crucial for the LASCAD™ analysis. Therefore, the lines with z-plane and voxel data from the different N text files were copied to replace the empty ones in the text file from one of the detectors. In Figure 8, the data from the detectors of rods 2 to 11 were used to replace the corresponding lines in the text file from rod 1 detector, since it already contained the absorbed pump flux/volume data of the Nd:YAG core in the detector volume's first portion. This helped in maintaining the same structure of a typical file directly exported from a detector volume in Zemax®. This single text file, with all the information of the core-doped Nd:YAG rods, was then imported into LASCAD™.

This resulted in a more compact laser resonator in LASCAD™ composed of the HR and PR 1064 nm mirrors, whose optical axes were aligned with that of the inclined laser rods 1 and 11, respectively, and a single laser rod with length equal to the sum of the lengths of each core-doped Nd:YAG rod of the array, as illustrated in Figure 9a for the merged laser beam of the N = 11 array in row 9. The total absorbed solar pump power for this single rod corresponded to the sum of the total pump power absorbed by each of the 11 rods. It is worth noting that the numerical analysis of the merged laser beams took into consideration the laser beam transmission losses (absorption, scattering, and diffraction) through both the Nd:YAG cores and the cooling water, as well as imperfections in the AR coatings on the end faces of the laser media and the HR coatings on the folding mirrors. For example, total round-trip loss of 32.2% was calculated for the merged TEM_{00}-mode laser beam from the N = 11 array in row 9, as shown in Figure 9a, resulting in maximum solar laser power of 153.69 W in fundamental mode regime (Figure 9b). $M_x^2 = 1.12$, $M_y^2 = 1.01$ laser beam quality factors were numerically found for this case. By using the beam propagation method in

LASCAD™, a near-Gaussian profile (Figure 9b) with beam waist radius of about 70.5 μm (Figure 9c) was numerically obtained at the PR 1064 nm output mirror, after several laser cavity iterations.

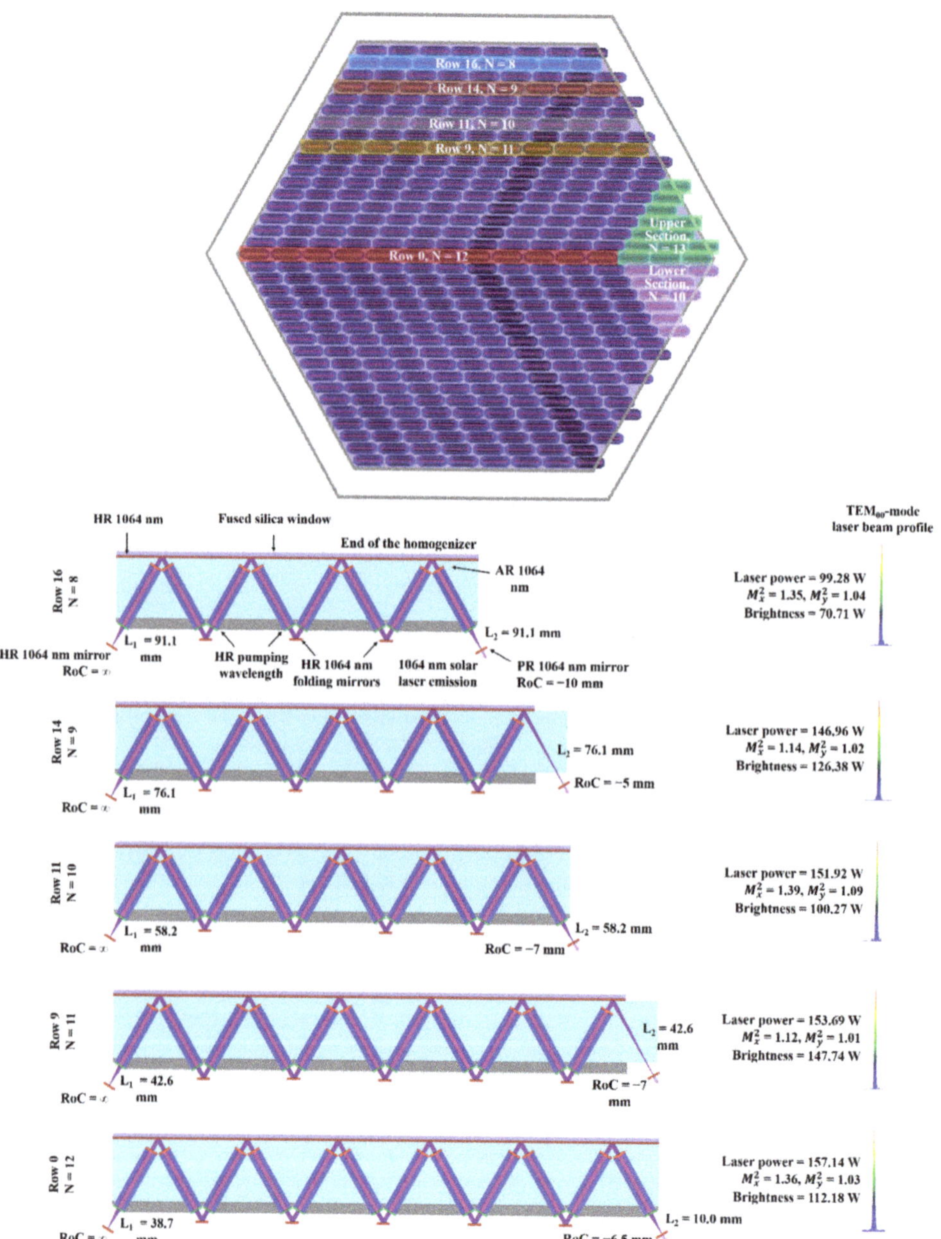

Figure 7. Illustration of the zigzag laser beam merging technique and TEM$_{00}$-mode laser performance for arrays of different N rods. L$_1$ is the separation length between the outer HR 1064 nm mirror and the lower end face of the leftmost rod, while L$_2$ represents the separation length between the lower or upper end face (determined by N) of the rightmost rod and the PR 1064 nm mirror. These separation lengths vary depending on N and the position of the array.

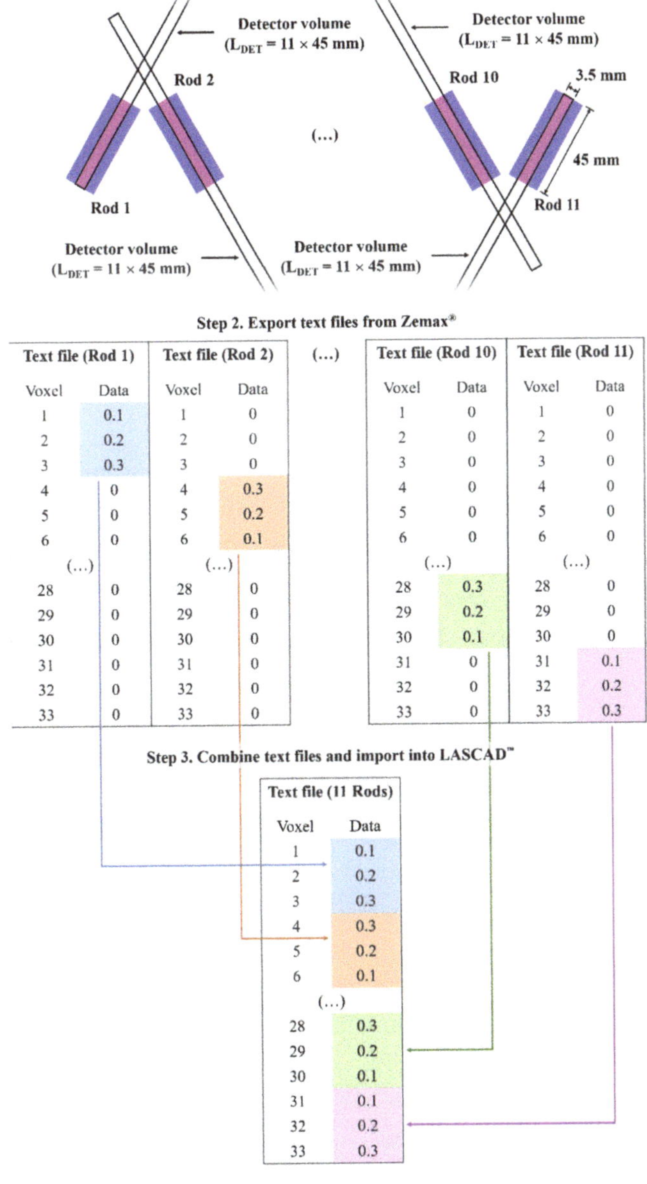

Figure 8. Simplified diagram of the process of combination of text files exported from Zemax® for LASCAD™ analysis of the extraction of a merged laser beam from an array with 11 rods. In this configuration, the HR and PR 1064 nm mirrors are closer to rod 1 and 11, respectively. The data presented in this diagram are for illustrative purposes only.

An analysis of the thermal induced effects on the N = 11 array of row 9 is presented in Figure 10. It shows that the maximum heat load (0.16 W/mm^3) and temperature (313 K) did not increase substantially in relation to the values observed with only a single Nd:YAG core.

The stress intensity (20.35 N/mm^2) rose over two times from the previously mentioned value, albeit still being considerably below the stress fracture limit of the Nd:YAG material.

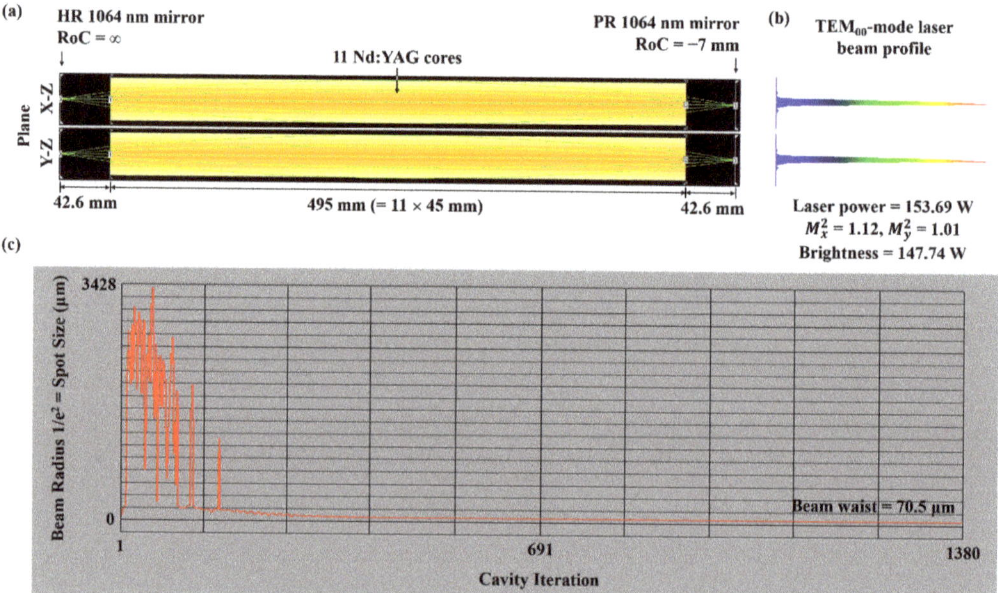

Figure 9. (**a**) Laser resonator design in LASCAD™ for the extraction of a merged TEM$_{00}$-mode laser beam from the N = 11 array of row 9, with (**b**) the respective beam profile. (**c**) Laser beam waist radius obtained at the PR 1064 nm output mirror, through the LASCAD™ beam propagation method.

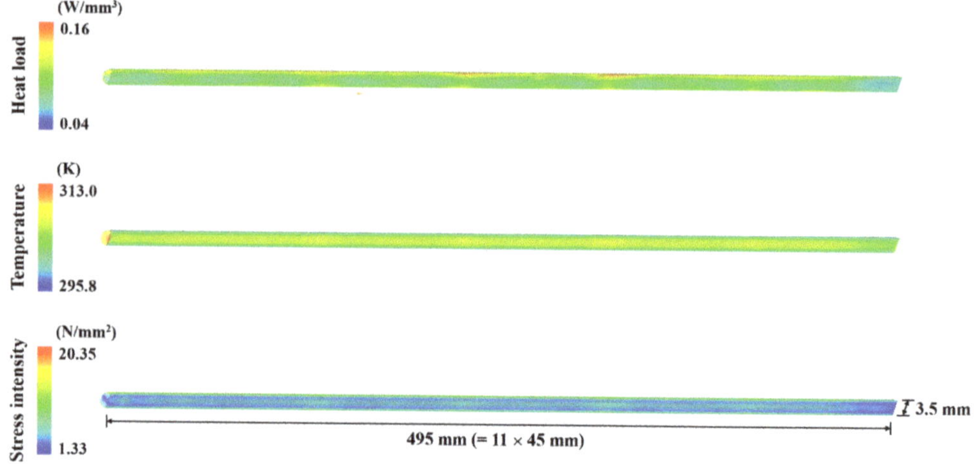

Figure 10. LASCAD™ analysis of heat load, temperature and stress intensity distribution in the N = 11 array of row 9.

The TEM$_{00}$-mode laser performance for five arrays of different N is detailed in Figure 7. With the laser beam merging technique, Gaussian laser beams of good quality with high TEM$_{00}$-mode power, and consequently, high brightness were attained, since it provided a better overlap between the pump and the fundamental mode volumes. Out of those five arrays, the N = 8 array of row 16 contributed with the lowest TEM$_{00}$-mode power of 99.28 W, which is 11.71 times greater than the 8.48 W total power from eight individual

TEM$_{00}$-mode laser beams (8 × 1.06 W). With the N = 12 array of row 0, the highest value of 157.14 W was obtained, corresponding to an increase of 12.35 times over the 12.72 W total power from twelve individual TEM$_{00}$-mode laser beams. The greatest enhancement, however, was achieved with the N = 9 array of row 14, which is 15.40 times higher than the 9.74 W total power from nine individual TEM$_{00}$-mode laser beams. The highest brightness figure of merit of 147.74 W was attained with the N = 11 array of row 9, which corresponds to a 182-fold increase over that from a single core-doped Nd:YAG rod. It is also 23 times higher than the experimental record of 6.46 W [16].

The TEM$_{00}$-mode and multimode laser power numerically extracted from each of the 37 arrays are presented in Figure 11. Despite the nearly uniform profile at the output end face of the homogenizer, a small difference in TEM$_{00}$-mode and multimode laser power from arrays with equal N was observed as a consequence of the position of certain rods that could not be placed in their entirety directly below the homogenizer's output face. Nevertheless, 5.2 kW total TEM$_{00}$-mode laser power may be extracted from 37 merged laser beams, corresponding to an improvement of three times in relation to the 1.8 kW attained before beam merging. Total multimode laser power of 8.6 kW was also determined.

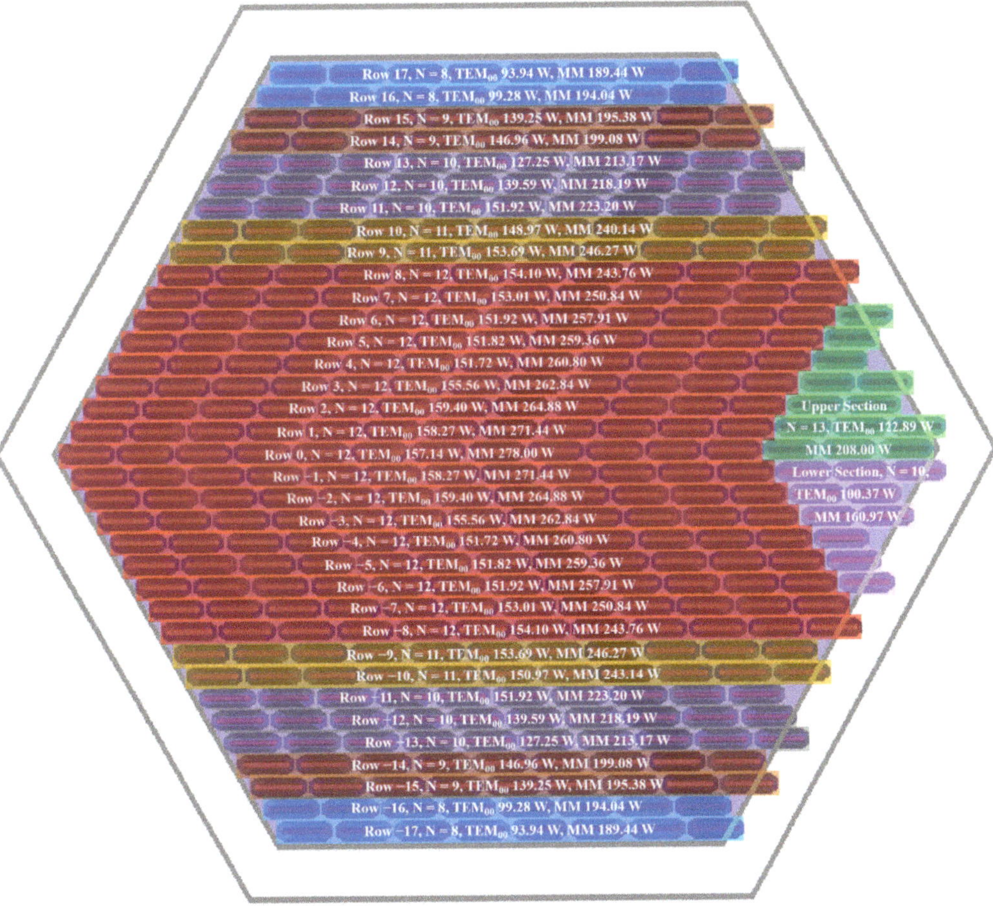

Figure 11. Summary of maximum TEM$_{00}$-mode (TEM$_{00}$) and multimode (MM) power attained from each laser beam extracted from different arrays of solar laser rods through the zigzag laser beam merging technique.

5. Conclusions

A multirod solar laser station concept was introduced to enable the production of high TEM_{00}-mode solar laser output power and beam brightness in MWSF. It was composed of a hexagonal shaped solar flux homogenizer with a fused silica window at its output end face, and 1657 core-doped Nd:YAG rods surrounded by a hexagonal reflector. The presence of the homogenizer facilitated the distribution of nearly the same amount of solar power to pump each rod and the substantial mitigation of the thermal lensing effects on each one, to the point that they were much less pronounced than in [38]. This facilitated the use of thin rods and long resonator cavities that are necessary to efficiently extract TEM_{00}-mode solar laser beams. With 3.5 mm diameter, 45 mm length Nd:YAG cores, a total laser power of 1.8 kW was numerically achieved.

The rod tilting and distance from the fused silica window may allow the possibility of employing a novel zigzag laser beam merging technique to boost the total TEM_{00}-mode solar laser output power and the beam brightness even further, despite having to reduce the number of rods from 1657 to 399. Not only was the TEM_{00}-mode power improved by three times, to 5.2 kW, from 37 merged laser beams, but the highest solar laser beam brightness figure of merit of about 148 W was also numerically determined, corresponding to a radical improvement of 23 times in relation to the previous experimental record [16]. The extraction of several high-power Gaussian TEM_{00}-mode solar laser beams of good quality and, thus, high brightness could greatly decrease the production costs with high-power lasers, paving the way for the future application of this renewable energy system in the industrial sector, as well as its expansion worldwide.

Author Contributions: Conceptualization, D.L., H.C., and J.A.; methodology, H.C., D.L., J.A. and C.R.V.; software, H.C., J.A., D.L., M.C. and B.D.T; validation, H.C., J.A. and D.L.; formal analysis, H.C., J.A., D.L. and D.G.; investigation, H.C., J.A. and D.L.; resources, D.L. and J.A.; data curation, H.C., J.A. and D.L.; writing—original draft preparation, H.C.; writing—review and editing, H.C., J.A., D.L., M.C., D.G., B.D.T. and C.R.V.; visualization, H.C.; supervision, D.L. and J.A.; project administration, D.L.; funding acquisition, D.L. All authors have read and agreed to the published version of the manuscript.

Funding: This research was funded by Fundação para a Ciência e a Tecnologia–Ministério da Ciência, Tecnologia e Ensino Superior, grant number UIDB/00068/2020.

Institutional Review Board Statement: Not applicable.

Informed Consent Statement: Not applicable.

Data Availability Statement: Not applicable.

Acknowledgments: The contract CEECIND/03081/2017 and the fellowship grants SFRH/BD/145322/2019, PD/BD/142827/2018, PD/BD/128267/2016 and SFRH/BPD/125116/2016 of Joana Almeida, Miguel Catela, Dário Garcia, Bruno D. Tibúrcio and Cláudia R. Vistas, respectively, are acknowledged.

Conflicts of Interest: The authors declare no conflict of interest. The funders had no role in the design of the study; in the collection, analyses, or interpretation of data; in the writing of the manuscript, or in the decision to publish the results.

References

1. Lando, M.; Kagan, J.A.; Shimony, Y.; Kalisky, Y.Y.; Noter, Y.; Yogev, A.; Rotman, S.R.; Rosenwaks, S. Solar-pumped solid state laser program. In Proceedings of the 10th Meeting on Optical Engineering in Israel, Jerusalem, Israel, 22 September 1997; Shladov, I., Rotman, S.R., Eds.; International Society for Optics and Photonics: Bellingham, WA, USA, 1997; Volume 3110, p. 196. [CrossRef]
2. Hemmati, H.; Biswas, A.; Djordjevic, I.B. Deep-Space Optical Communications: Future Perspectives and Applications. *Proc. IEEE* **2011**, *99*, 2020–2039. [CrossRef]
3. Rather, J.D.G.; Gerry, E.T.; Zeiders, G.W. Investigation of possibilities for solar powered high energy lasers in space. *NASA Tech. Rep. Serv.* **1977**. Available online: https://ntrs.nasa.gov/search.jsp?R=19770019534 (accessed on 15 March 2020).
4. Vasile, M.; Maddock, C.A. Design of a formation of solar pumped lasers for asteroid deflection. *Adv. Sp. Res.* **2012**, *50*, 891–905. [CrossRef]

5. Yabe, T.; Uchida, S.; Ikuta, K.; Yoshida, K.; Baasandash, C.; Mohamed, M.S.; Sakurai, Y.; Ogata, Y.; Tuji, M.; Mori, Y.; et al. Demonstrated fossil-fuel-free energy cycle using magnesium and laser. *Appl. Phys. Lett.* **2006**, *89*, 261107. [CrossRef]

6. Graydon, O. A sunny solution. *Nat. Photonics* **2007**, *1*, 495–496. [CrossRef]

7. Oliveira, M.; Liang, D.; Almeida, J.; Vistas, C.R.; Gonçalves, F.; Martins, R. A path to renewable Mg reduction from MgO by a continuous-wave Cr:Nd:YAG ceramic solar laser. *Sol. Energy Mater. Sol. Cells* **2016**, *155*, 430–435. [CrossRef]

8. Kiss, Z.J.; Lewis, H.R.; Duncan, R.C., Jr. Sun Pumped Continuous Optical Maser. *Appl. Phys. Lett.* **1963**, *2*, 93–94. [CrossRef]

9. Lando, M.; Kagan, J.; Linyekin, B.; Dobrusin, V. A solar-pumped Nd:YAG laser in the high collection efficiency regime. *Opt. Commun.* **2003**, *222*, 371–381. [CrossRef]

10. Yabe, T.; Ohkubo, T.; Uchida, S.; Yoshida, K.; Nakatsuka, M.; Funatsu, T.; Mabuti, A.; Oyama, A.; Nakagawa, K.; Oishi, T.; et al. High-efficiency and economical solar-energy-pumped laser with Fresnel lens and chromium codoped laser medium. *Appl. Phys. Lett.* **2007**, *90*, 261120. [CrossRef]

11. Liang, D.; Almeida, J. Highly efficient solar-pumped Nd:YAG laser. *Opt. Express* **2011**, *19*, 26399. [CrossRef]

12. Dinh, T.H.; Ohkubo, T.; Yabe, T.; Kuboyama, H. 120 watt continuous wave solar-pumped laser with a liquid light-guide lens and an Nd:YAG rod. *Opt. Lett.* **2012**, *37*, 2670. [CrossRef]

13. Xu, P.; Yang, S.; Zhao, C.; Guan, Z.; Wang, H.; Zhang, Y.; Zhang, H.; He, T. High-efficiency solar-pumped laser with a grooved Nd:YAG rod. *Appl. Opt.* **2014**, *53*, 3941. [CrossRef]

14. Guan, Z.; Zhao, C.; Li, J.; He, D.; Zhang, H. 32.1 W/m^2 continuous wave solar-pumped laser with a bonding Nd:YAG/YAG rod and a Fresnel lens. *Opt. Laser Technol.* **2018**, *107*, 158–161. [CrossRef]

15. Liang, D.; Vistas, C.R.; Tibúrcio, B.D.; Almeida, J. Solar-pumped Cr:Nd:YAG ceramic laser with 6.7% slope efficiency. *Sol. Energy Mater. Sol. Cells* **2018**, *185*, 75–79. [CrossRef]

16. Liang, D.; Almeida, J.; Vistas, C.R.; Guillot, E. Solar-pumped Nd:YAG laser with 31.5 W/m^2 multimode and 7.9 W/m^2 TEM$_{00}$-mode collection efficiencies. *Sol. Energy Mater. Sol. Cells* **2017**, *159*, 435–439. [CrossRef]

17. Hwang, I.H.; Lee, J.H. Efficiency and threshold pump intensity of CW solar-pumped solid-state lasers. *IEEE J. Quantum Electron.* **1991**, *27*, 2129–2134. [CrossRef]

18. Koechner, W. *Solid-State Laser Engineering*, 5th ed.; Springer Series in Optical Sciences; Springer: Berlin/Heidelberg, Germany, 1999; ISBN 978-3-662-14221-9.

19. Clarkson, W.A. Thermal effects and their mitigation in end-pumped solid-state lasers. *J. Phys. D. Appl. Phys.* **2001**, *34*, 2381–2395. [CrossRef]

20. Lupei, V.; Lupei, A.; Gheorghe, C.; Ikesue, A. Emission sensitization processes involving Nd^{3+} in YAG. *J. Lumin.* **2016**, *170*, 594–601. [CrossRef]

21. Weksler, M.; Shwartz, J. Solar-pumped solid-state lasers. *IEEE J. Quantum Electron.* **1988**, *24*, 1222–1228. [CrossRef]

22. Ostermeyer, M.; Brandenburg, I. Simulation of the extraction of near diffraction limited Gaussian beams from side pumped core doped ceramic Nd:YAG and conventional laser rods. *Opt. Express* **2005**, *13*, 10145. [CrossRef]

23. Sträßer, A.; Ostermeyer, M. Improving the brightness of side pumped power amplifiers by using core doped ceramic rods. *Opt. Express* **2006**, *14*, 6687. [CrossRef]

24. Almeida, J.; Liang, D. Design of a high brightness solar-pumped laser by light-guides. *Opt. Commun.* **2012**, *285*, 5327–5333. [CrossRef]

25. Liang, D.; Almeida, J. Design of ultrahigh brightness solar-pumped disk laser. *Appl. Opt.* **2012**, *51*, 6382. [CrossRef]

26. Liang, D.; Almeida, J. Multi-Fresnel lenses pumping approach for improving high-power Nd:YAG solar laser beam quality. *Appl. Opt.* **2013**, *52*, 5123. [CrossRef]

27. Garcia, D.; Liang, D.; Almeida, J. Core-doped Nd:YAG disk solar laser uniformly pumped by six Fresnel lenses. In Proceedings of the 8th Iberoamerican Optics Meeting and 11th Latin American Meeting on Optics, Lasers, and Applications, Porto, Portugal, 18 November 2013; Martins Costa, M.F.P.C., Ed.; International Society for Optics and Photonics: Bellingham, WA, USA, 2013; Volume 8785, p. 87850W. [CrossRef]

28. Strite, T.; Gusenko, A.; Grupp, M.; Hoult, T. Multiple Laser Beam Material Processing. *Biul. Inst. Spaw.* **2016**, *60*, 30–33. [CrossRef]

29. Eifel, S.; Holtkamp, J. Multi-Beam Technology Boosts Cost Efficiency. Available online: https://www.industrial-lasers.com/micromachining/article/16485583/multibeam-technology-boosts-cost-efficiency (accessed on 17 May 2020).

30. Gillner, A.; Finger, J.; Gretzki, P.; Niessen, M.; Bartels, T.; Reininghaus, M. High Power Laser Processing with Ultrafast and Multi-Parallel Beams. *J. Laser Micro/Nanoeng.* **2019**, *14*, 129–137. [CrossRef]

31. Olsen, F.O.; Hansen, K.S.; Nielsen, J.S. Multibeam fiber laser cutting. *J. Laser Appl.* **2009**, *21*, 133–138. [CrossRef]

32. Fazilov, A.; Riskiev, T.T.; Abdurakhmanov, A.A.; Bakhramov, S.A.; Makhkamov, S.; Mansurov, M.M.; Mukhamediev, E.J.; Paiziev, S.D.; Klychev, S.I.; Saribaev, A.S.; et al. Concentrated solar energy conversion to powerful laser radiation on neodymium activated yttrium-aluminum garnet. *Appl. Sol. Energy* **2008**, *44*, 93–96. [CrossRef]

33. Tibúrcio, B.D.; Liang, D.; Almeida, J.; Garcia, D.; Vistas, C.R. Dual-rod pumping concept for TEM$_{00}$-mode solar lasers. *Appl. Opt.* **2019**, *58*, 3438. [CrossRef] [PubMed]

34. Almeida, J.; Liang, D.; Tibúrcio, B.D.; Garcia, D.; Vistas, C.R. Numerical modeling of a four-rod pumping scheme for improving TEM$_{00}$-mode solar laser performance. *J. Photonics Energy* **2019**, *9*, 1. [CrossRef]

35. Liang, D.; Almeida, J.; Garcia, D.; Tibúrcio, B.D.; Guillot, E.; Vistas, C.R. Simultaneous solar laser emissions from three Nd:YAG rods within a single pump cavity. *Sol. Energy* **2020**, *199*, 192–197. [CrossRef]

36. Costa, H.; Almeida, J.; Liang, D.; Garcia, D.; Catela, M.; Tibúrcio, B.D.; Vistas, C.R. Design of a multibeam solar laser station for a megawatt solar furnace. *Opt. Eng.* **2020**, *59*, 086103. [CrossRef]
37. Almeida, J.; Liang, D.; Costa, H.; Garcia, D.; Tibúrcio, B.D.; Catela, M.; Vistas, C.R. Seven-rod pumping concept for simultaneous emission of seven TEM_{00}-mode solar laser beams. *J. Photonics Energy* **2020**, *10*, 038001. [CrossRef]
38. Costa, H.; Almeida, J.; Liang, D.; Tibúrcio, B.D.; Garcia, D.; Catela, M.; Vistas, C.R. Quasi-Gaussian Multibeam Solar Laser Station for a Megawatt Solar Furnace. *J. Sol. Energy Res. Updat.* **2021**, *8*, 11–20. [CrossRef]
39. Liang, D.; Almeida, J.; Tibúrcio, B.D.; Catela, M.; Garcia, D.; Costa, H.; Vistas, C.R. Seven-Rod Pumping Approach for the Most Efficient Production of TEM_{00} Mode Solar Laser Power by a Fresnel Lens. *J. Sol. Energy Eng.* **2021**, *143*, 061004. [CrossRef]
40. Liang, D.; Almeida, J. Solar-Pumped TEM_{00} Mode Nd:YAG laser. *Opt. Express* **2013**, *21*, 25107. [CrossRef] [PubMed]
41. Liang, D.; Almeida, J.; Vistas, C.R.; Guillot, E. Solar-pumped TEM_{00} mode Nd:YAG laser by a heliostat-Parabolic mirror system. *Sol. Energy Mater. Sol. Cells* **2015**, *134*, 305–308. [CrossRef]
42. Almeida, J.; Liang, D.; Vistas, C.R.; Bouadjemine, R.; Guillot, E. 5.5 W continuous-wave TEM_{00}-mode Nd:YAG solar laser by a light-guide/2V-shaped pump cavity. *Appl. Phys. B* **2015**, *121*, 473–482. [CrossRef]
43. Vistas, C.R.; Liang, D.; Almeida, J. Solar-pumped TEM_{00} mode laser simple design with a grooved Nd:YAG rod. *Sol. Energy* **2015**, *122*, 1325–1333. [CrossRef]
44. Liang, D.; Almeida, J.; Vistas, C.R.; Oliveira, M.; Gonçalves, F.; Guillot, E. High-efficiency solar-pumped TEM_{00}-mode Nd:YAG laser. *Sol. Energy Mater. Sol. Cells* **2016**, *145*, 397–402. [CrossRef]
45. Duocastella, M.; Arnold, C.B. Bessel and annular beams for materials processing. *Laser Photonics Rev.* **2012**, *6*, 607–621. [CrossRef]
46. Catela, M.; Liang, D.; Vistas, C.R.; Garcia, D.; Tibúrcio, B.D.; Costa, H.; Almeida, J. Six-rod/six-beam concept for revitalizing TEM_{00} mode lamp-pumped lasers. *Opt. Eng.* **2020**, *59*, 126108. [CrossRef]
47. Guillot, E.; Rodriguez, R.; Boullet, N.; Sans, J.-L. Some details about the third rejuvenation of the 1000 kWth solar furnace in Odeillo: Extreme performance heliostats. In Proceedings of the AIP Conference Proceedings; 2018; Volume 2033, p. 040016. Available online: https://aip.scitation.org/doi/abs/10.1063/1.5067052 (accessed on 24 August 2021). [CrossRef]
48. SFERA Access to PROMES Facilities. Available online: http://sfera.sollab.eu/index.php?page=access_promes (accessed on 6 May 2020).
49. ASTM G173-03(2012). *Standard Tables for Reference Solar Spectral Irradiances: Direct Normal and Hemispherical on 37° Tilted Surface*; ASTM: West Conshohocken, PA, USA, 2012. [CrossRef]
50. Zhao, B.; Changming, Z.; He, J.; Yang, S. The study of active medium for solar-pumped solid-state lasers. *Acta Opt. Sin.* **2007**, *27*, 1797–1801.
51. Almeida, J.; Liang, D.; Vistas, C.R.; Guillot, E. Highly efficient end-side-pumped Nd:YAG solar laser by a heliostat–parabolic mirror system. *Appl. Opt.* **2015**, *54*, 1970. [CrossRef] [PubMed]

MDPI

St. Alban-Anlage 66

4052 Basel

Switzerland

Tel. +41 61 683 77 34

Fax +41 61 302 89 18

www.mdpi.com

Energies Editorial Office

E-mail: energies@mdpi.com

www.mdpi.com/journal/energies

9 783036 560373